# Monoclonal Antibodies: A Manual of Techniques

Author

Heddy Zola
Department of Clinical Immunology
Flinders Medical Centre
and
Flinders University of South Australia
Bedford Park,
Australia

**CRC Press, Inc.**
**Boca Raton, Florida**

**Library of Congress Cataloging-in-Publication Data**

Zola, Heddy.
Monoclonal antibodies.

Includes bibliographies and index.
1. Antibodies, Monoclonal. 2. Immunology—Technique.
I. Title. [DNLM: 1. Antibodies, Monoclonal.
2. Hybridomas. 3. Immunologic Technics. QW 575 Z86m]
QR186.85.Z65 1987 616.07'93 86-18394
ISBN 0-8493-6476-0



Direct all inquiries to CRC Press, Inc., 2000 Corporate Blvd., N.W., Boca Raton, Florida, 33431.


Second Printing, 1988
Third Printing, 1989

International Standard Book Number 0-8493-6476-0

Library of Congress Card Number 86-18394
Printed in the United States

# PREFACE

In the past 10 years since the original description of antibody production by a somatic cell hybrid line, monoclonal antibodies have become standard reagents in many biological applications. There have been few major changes in the techniques used to make and to utilize monoclonal antibodies, and the mechanism of hybridization is still not understood in any detail. However, a few years of experience with the techniques have brought about important refinements, and have established a body of data which allows choices to be made between some of the alternative procedures. The production and use of monoclonal antibodies can, as a result, proceed in a fairly predictable way, provided that the best techniques are chosen for the particular objectives of the project. This book describes, in detail, techniques for the production and use of monoclonal antibodies. Methods are set out in the form of experimental protocols, so that the book can serve as a laboratory manual. The rationale behind each method is discussed, where appropriate. Where several alternative methods are available, a rational basis is provided, where possible, for the choice of methods appropriate to the reader's own needs. The book does not set out to review specific fields of use, however, since it is intended to cover in detail those aspects of interest to all scientists working with monoclonal antibodies.

ACKNOWLEDGMENTS

This book is principally concerned with methodology. The techniques described have been used in projects which I have either directed or participated in. However, many of my colleagues have contributed in important ways, either with ideas or by doing the actual work. Current or former higher-degree students (Doug Brooks, Ian Beckman, and Russell Hogg), research assistants (Peter McNamara, Ian Hunter, Helen Moore, Joe Webster, and Angela Potter), and Hospital Scientists (Miriam Thomas, Pete Macardle, and Virginia Griffith) have done much of the work which serves as a base for this book. Scientific colleagues who have contributed ideas and suggestions, and have taught me some of the methods, include Drs. Anthony Hodgson, Sim Hee Neoh, Keryn Williams, John Bradley, and Art Hohmann. Professor John Bradley, Head of the Department of Clinical Immunology, has provided an environment which made the research possible, and has given active encouragement and support to research in his Department.

I prepared the early draft of this book on an Apple (and an Apple look-alike) and I owe a real debt to Messrs. Jobs and Wozniak. Writing books must have been such hard work in the days before microcomputers for the masses. My drafts, together with a multitude of alterations, were converted to the finished manuscript by Mary Brown, whose diligence and careful work is greatly appreciated. The art-work was produced by Dennis Jones of the Media and Illustration Unit, Flinders Medical Centre, and by Adrian Wright and Angela Potter. The photographs are my own, but my colleagues had to wait patiently with syringe or scalpel poised, while I fiddled with the focus or the flash.

Dr. Anthony Hodgson read a later draft and provided valuable criticism, particularly of the sections dealing with immunohistochemistry, an area in which his expertise is much greater than mine. Dr. Sim Neoh provided details for the immunoglobulin purification methods of Chapter 4. Dr. Keryn Williams read the whole book in draft form, and provided extensive and valuable critical comments on the scientific content, the clarity, and the English. It is all too easy to write text which is quite clear to the writer, and totally unclear to the reader. If this book turns out to be readable, this will be due, in large measure, to Keryn Williams' constructive criticism. Having sought the help and advice of my colleagues, I chose to ignore some of their comments. The responsibility for any residual flaws is therefore squarely mine.

## THE AUTHOR

Heddy Zola is Chief Hospital Scientist in the Department of Clinical Immunology, Flinders Medical Centre, and Reader in Immunology in the Flinders University of South Australia. He received a Bachelor of Science degree in Chemistry from the University of Birmingham, in England, and a Doctor of Philosophy degree in Biophysics, from the University of Leeds, England.

Following his Ph.D. studies Dr. Zola worked in the Biophysics Department of Leeds University, before joining the Wellcome Research Laboratories, Beckenham, Kent, where he carried out biophysical and biochemical studies on a variety of biologically active proteins. Involvement in a project aimed at the preparation of antilymphocyte serum, for the use in the treatment and prevention of transplant rejection in man, led to an interest in the immunology of the lymphocyte surface. When Kohler and Milstein described the first hybridoma, the potential of this technique as a tool in studies of cell membrane molecules soon became apparent. Dr. Zola started to make monoclonal antibodies against human lymphocyte antigens, initially at Wellcome and subsequently in Australia.

Dr. Zola was born and grew up in Africa, where his parents had settled as refugees from Europe. After attending schools in Zanzibar and Kenya, he went to England in 1959, initially to study, married and stayed on in England until 1978. The lure of sunshine and opens spaces eventually proved irresistible, and Dr. Zola, his wife Marion, and their two daughters moved to Australia, where they live in the city of Adelaide.

# TABLE OF CONTENTS

Chapter 3
Making Hybridomas ........................................................ 23

Chapter 5

Using Monoclonal Antibodies: Cell and Tissue Markers....................................89

Chapter 1

# INTRODUCTION

## I. AIM AND SCOPE OF THIS BOOK

The first publication describing the preparation of a monoclonal antibody by somatic cell hybridization appeared in 1975.[1] Less than a decade later, in 1984, the authors of that paper (Kohler and Milstein) have shared a Nobel Prize, and monoclonal antibodies are regularly mentioned in the lay press. The layman really does not need to know much about monoclonal antibodies; they are high technology, the product of genetic engineering, used in making magic bullets, and quite definitely "a good thing."

On the other hand, biological and medical scientists in a variety of disciplines do need to know much more about monoclonal antibodies, and if they are not immunologists (and, perhaps, even if they are) they may need some help. Monoclonal antibodies are available for use, or at least talked about, in a variety of fields, from cancer diagnosis to the food industry; from transplant surgery to plant pathology. The scientist in any biological or medical field may need to answer these questions:

- Do I need monoclonals?
- Do I need to make them myself?
- How do I use them?
- What can I realistically expect from them?

This book is intended to enable biologists and medical scientists to answer such questions for themselves, in the context of their own particular needs. The book provides detailed methodology to enable readers to make their own antibodies and to use monoclonal antibodies to their best advantage.

## II. ANTIBODY AS A SPECIFIC PROBE

Antibodies are proteins made by many animal species as part of the specific immune response to foreign substances. Specificity is the hallmark of the immune response; recovery from measles leads to lifelong immunity from measles, but confers no protection against chicken pox. One component of this specific response is the synthesis, by B lymphocytes and plasma cells, of antibody which reacts with a high affinity with the stimulating substance (the antigen), but not with other substances. This natural ability to make proteins with a high order of specificity in distinguishing different molecular structures can be used in a variety of ways. Antibody obtained by injecting a rabbit with human insulin reacts specifically with human insulin and can be used as a probe for the hormone. The antibody can serve as the basis for a quantitative assay for the hormone; it can be used to detect cells secreting the hormone in tissue sections; it can be used to purify the hormone from a complex mixture of proteins. Thus, antibodies have long been used as reagents in the detection, measurement, and purification of biological molecules. Antibodies have also been used for treatment; antibody made in horses against bacterial toxins or snake venoms provides emergency treatment, albeit with considerable risk of an adverse immune reaction to the horse antibody, itself seen as a foreign molecule by the immune system of the patient.

Until recently such antibodies were obtained by immunizing an animal and then bleeding it. The antibodies are secreted into the serum, and the protein fraction con-

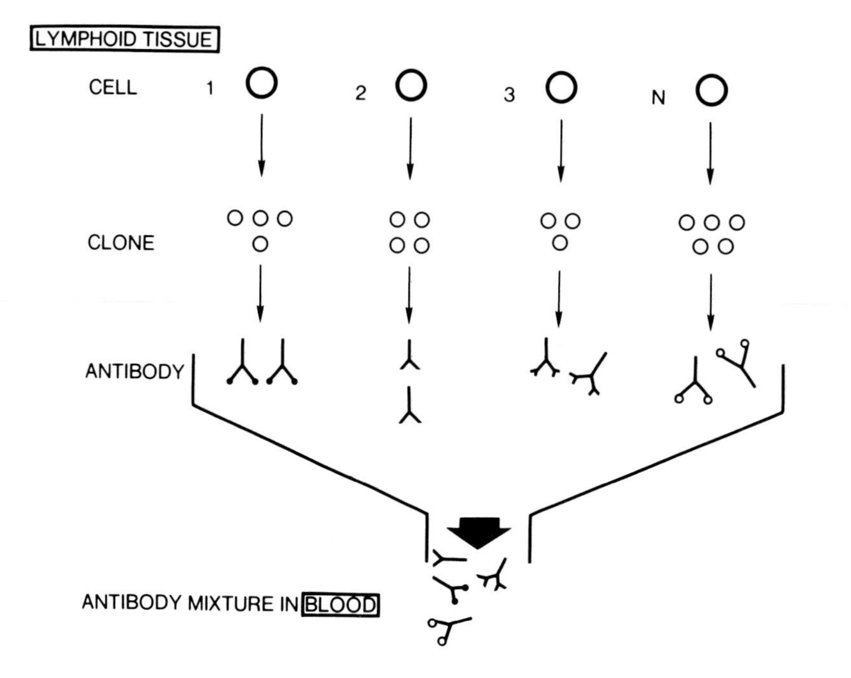


FIGURE 1. The immune response in the animal: B lymphocytes respond to different antigenic determinants (top line). Each stimulated cell multiplies, giving rise to a clone of cells; each clone produces one antibody specificity. The secreted antibody is, however, mixed before we can obtain it from the blood.

taining antibody activity (immunoglobulin) can be purified by straightforward biochemical means. However, the final product is not pure antibody against the immunizing antigen. All antibodies, against whatever antigen, have a common structure (with some variations, which will be discussed later). Thus, antibody against the venom of the Australian tiger snake and antibody against measles virus will have the same molecular weight and overlapping distributions of isoelectric points; they will be impossible to separate except by methods based on the affinity of the antibody for its specific antigen. Two antibodies against different antigens may differ in amino acid sequence in only a few small ("hypervariable") regions of the protein. Because all animals are constantly subject to environmental antigenic stimulation and maintain a relatively constant concentration of immunoglobulin in the serum, the effect of intentional immunization is to add some additional immunoglobulin to an already complex mixture; after vigorous immunization the immunologist can hope for no more than 1 to 2% of the total immunoglobulin to be directed against the immunogen. If the antigen is available in pure form and in quantity, the antibody can be isolated from this mixture by affinity-based separation methods. Very often the antigen is available neither pure nor in abundance.

The antibodies which are found in the serum have been secreted by lymphoid cells in various organs, including the bone marrow, spleen, and lymph nodes. In these organs, each individual cell is capable of secreting only one specificity of antibody. If, instead of letting these pure products mix in the serum, we could isolate a cell making antibody suited to our purpose and then clone that cell — make multiple copies of the cell all producing identical antibody — we would have a source of pure antibody. If the cells could be made self-renewing indefinitely, we would have essentially unlimited supplies of invariant antibody. This is illustrated in Figures 1 and 2. This concept is the essence of the monoclonal antibody method.

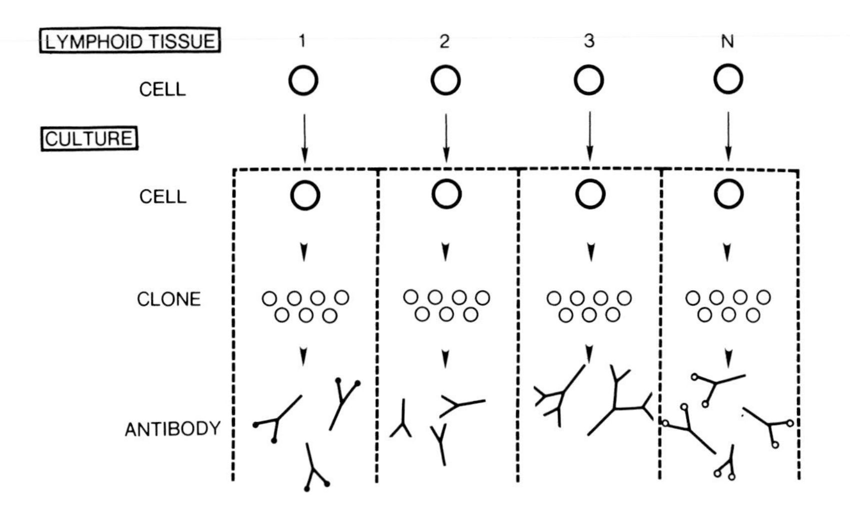


FIGURE 2. Keeping the cells separated yields pure antibody: if the stimulated B lymphocytes can be kept separated and made to form antibody-secreting clones in separate cultures in vitro, we can harvest pure antibody. Compare with Figure 1.

## III. MONOCLONAL ANTIBODY

Monoclonal antibodies are, thus, pure reagents; each molecule is identical to the rest in the preparation. They are produced by a culture of cells in which all the cells derive from a single cloned cell — a monoclonal population. The concept of monoclonality is not as clear as it may appear. You, the reader, are derived from a single cell, yet you are manifestly not monoclonal. In a cell culture derived by cloning we make the assumption that the cells are all identical, yet in reality they are not, as we shall see later. The important point is that any variation which may develop does not involve the antibody product. Thus, monoclonality is achieved by cloning the cells, but is best defined in terms of the antibody produced.

Figure 3 shows the electrophoretic homogeneity of a serum containing a monoclonal antibody, compared with the heterogenous immunoglobulin in a normal serum.

There are several ways of deriving monoclonal antibody-producing cell lines from a mixture of immunized lymphocytes. Two distinct changes have to be worked: the cells have to be immortalized, caused to multiply indefinitely; and they must be cloned — separated into individual pure cultures (Figure 2). The cloning is relatively straightforward — a matter of culturing cells so that they start and remain separated from each other. The immortalization can be achieved in principle by two methods: removal of the cells from constraints which prevent unlimited replication, and transformation by a virus or other source of genetic message that causes the cell to multiply indefinitely. In practice only the second approach has been used, and the transforming agent has been either virus (Epstein-Barr virus, which transforms human B lymphocytes) or the DNA from another cell, one that constitutively divides indefinitely. The DNA is normally supplied by hydridizing the two cells, but can be introduced directly into a cell, a process known as transfection.

The vast majority of useful monoclonal antibodies prepared to date have been made by hybridization, and this is the approach used in this book. The principles of the process are illustrated schematically in Figure 4, and the technical details of the process are described in Chapter 3.

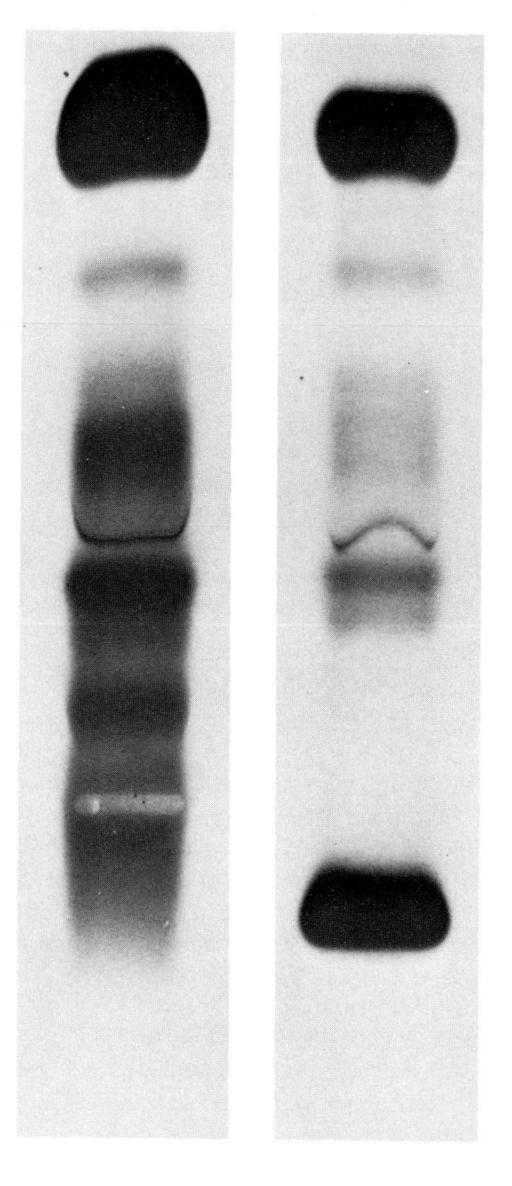


FIGURE 3. Serum protein electrophoresis of normal serum (lane 1) and serum from a patient with multiple myeloma (lane 2). Multiple myeloma is a malignant proliferation of a single clone of antibody-producing cells, and a sharp band of monoclonal immunoglobulin is seen. In normal serum the heterogeneity of the antibody molecules is reflected in a broad band of immunoglobulin molecules, all differing in composition. Sera from mice which have been injected with a laboratory-prepared hybridoma resemble the myeloma serum (see Chapter 4).

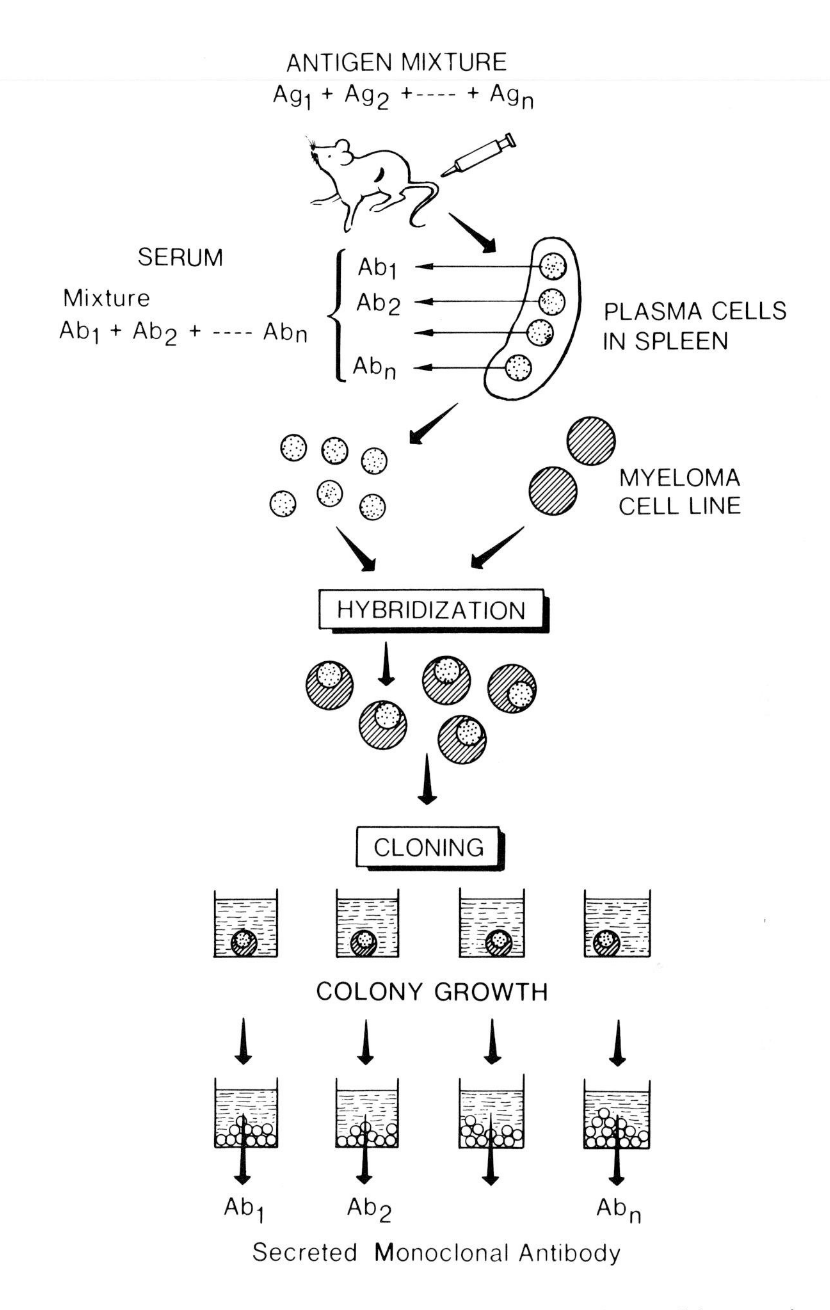


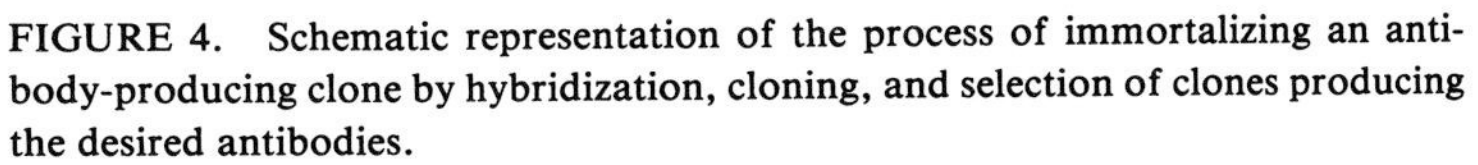
FIGURE 4. Schematic representation of the process of immortalizing an antibody-producing clone by hybridization, cloning, and selection of clones producing the desired antibodies.

## IV. MONOCLONAL AND POLYCLONAL ANTIBODY COMPARED

### A. One Antibody or Many Different Antibodies

A solution of monoclonal antibody contains a single antibody specificity and affinity and a single immunoglobulin isotype, whereas a polyclonal antibody preparation contains a variety of antibody molecules directed against the antigen, as well as antibodies which do not react with the antigen of interest. These differences between monoclonal and polyclonal antibody result in many practical differences in the behavior of the

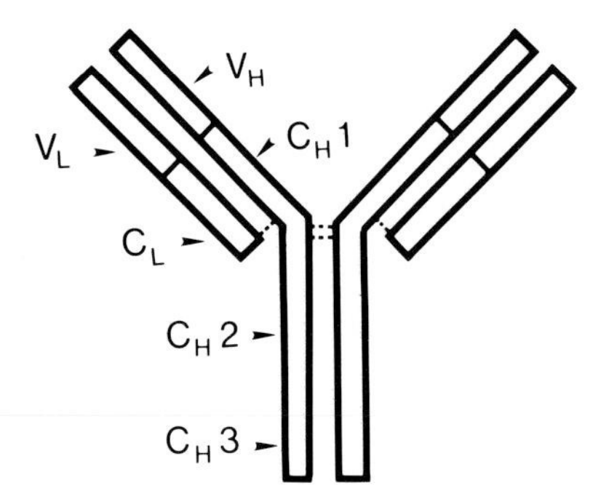


FIGURE 5. Structure of a monomeric immunoglobulin molecule. The molecule consists of light (L) and heavy (H) chains. Each L chain is composed of a variable ($V_L$) and a constant ($C_L$) domain. Each heavy chain is composed of a variable ($V_H$) and three constant ($C_H$1-3) domains. The antigen-binding site is made up by the juxtaposition of the $V_L$ and $V_H$ domains. Each molecule consists of two H-L pairs; monomeric immunoglobulin, thus, has two binding sites; it is divalent. Different immunoglobulin types (isotypes, $IgG_1$, $IgG_{2a}$, etc.) differ in the C domains but can have the same antigen-binding site.

antibodies. The differences usually make the monoclonal antibody the superior reagent, but this is not invariably so. The object of this section is to explain the differences in behavior between monoclonal and polyclonal antibodies and to derive from this information a rationale for making the best use of both monoclonal and polyclonal antibodies. Examples will be developed later on, particularly in Chapters 5 and 6, but the general principles can be considered here.

### B. Different Antibodies Against the Same Antigen

A polyclonal antibody preparation contains not only antibodies against "irrelevant" antigens, it contains multiple different antibodies against the antigen of interest. There will be antibodies of different isotypes (heavy chain types, see Figure 5 and 6 and Table 1), and these will behave differently in the way they interact with the antigen, and with cells, complement, and any other component of an experiment. There will be antibodies with minor differences in amino acid sequence, which react with the same antigen but with different affinity. The properties of the polyclonal antibody will be an average property, depending on the composition of the mixture. A monoclonal antibody will have a single isotype and a single affinity. This is not always advantageous, as will be seen below.

### C. Single or Multiple Antigenic Epitopes

The monospecificity of monoclonal antibodies prevents them from exhibiting those antibody functions that require the presence of antibodies against several determinants on the antigen. Thus, monoclonal antibodies do not generally precipitate antigens unless the determinant detected is present in multiple copies on the molecule. The reason is that a sufficiently large antigen-antibody lattice is not formed. However, mixtures of monoclonal antibodies against different determinants on the antigen molecule can precipitate effectively. Alternatively, the solubility of the small immune complex formed between monoclonal antibody and monovalent antigen may be reduced by the addition of reagents such as polyethylene glycol, leading to the formation of a precipitate.

### D. Recognition of Patterns Consisting of Several Antigenic Determinants

Some uses of antisera involve the recognition, not of a single antigen, but of a pat-

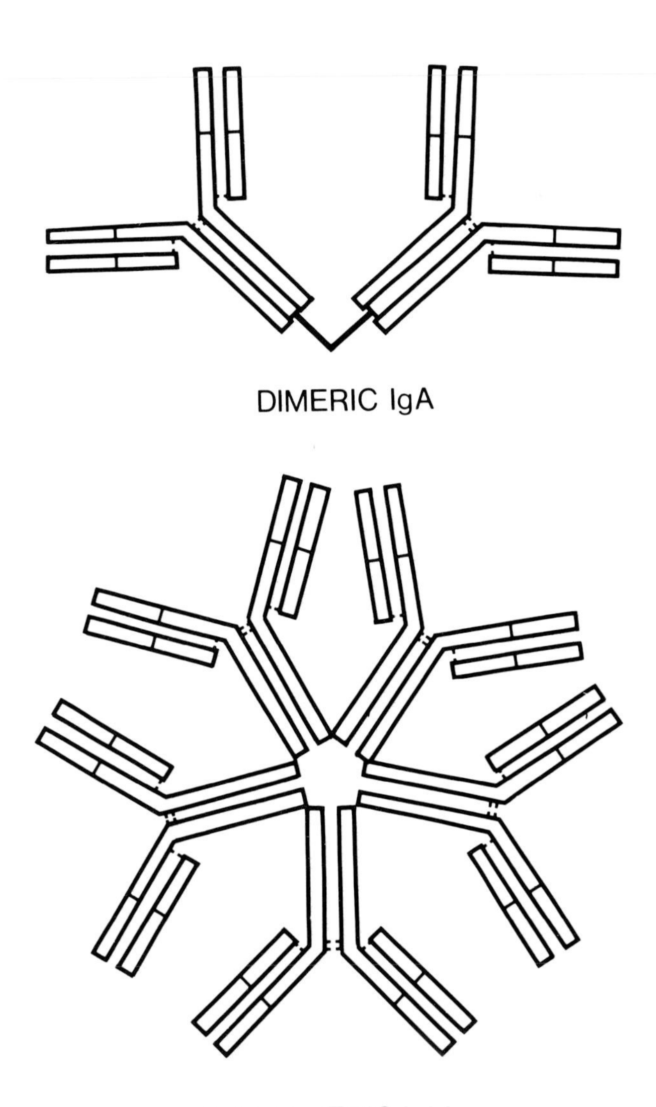


FIGURE 6. Structure of dimeric and pentameric immunoglobulins. These molecules are oligomers of the basic structure shown in Figure 5.

tern consisting of several antigens. An example is tissue typing. Current typing sera are alloantisera (antisera made by immunization within the species) and establish a pattern consisting of multiple antigenic differences. Monoclonal antibodies react with single determinants, which may be shared by a number of genetically different histocompatibility antigens. Thus, with a few exceptions monoclonal antibodies have failed to correlate with existing tissue types. An additional reason for the differences between mouse monoclonal antibodies and human typing sera is that the mouse and human will not necessarily respond immunologically to the same epitopes. The relative importance of these two different reasons for the lack of correlation between mouse monoclonal and human alloantisera in tissue typing will become clearer when human monoclonal antibodies against polymorphic tissue antigens become available.

Table 1
MOUSE IMMUNOGLOBULIN ISOTYPES

| | Isotype structure | Complement fixation | Comment |
|---|---|---|---|
| IgM | Pentamer + J chain | Yes | Produced in primary response; IgM in boosted animals often against carbohydrate |
| IgD | Dimer | No | Cell-bound antibody, IgD hybridomas very rare |
| IgG1 | Monomer | Weak or -ve | Frequent hybridoma product |
| IgG2a | Monomer | Yes | Frequent hybridoma product |
| IgG2b | Monomer | Yes | Least frequent IgG subclass |
| IgG3 | Monomer | Yes | Often directed against carbohydrate epitopes |
| IgE | Monomer | No | Special immunization needed to induce IgE hybridomas |
| IgA | Monomer or dimer + J chain; secretory piece | No (activates alternative pathway) | Minor serum component; major secretory immunoglobulin |

E. Sensitivity and Specificity

A consequence of the specificity of monoclonal antibodies is that relatively few antibody molecules bind to a target cell, thus, limiting sensitivity. This can be circumvented by using appropriate mixtures of monoclonal antibodies to increase binding. Although monoclonal antibodies tend to be less sensitive than polyclonal mixtures, they show much greater specificity and lower background staining.

F. Labile Epitopes

The various fixatives used in histology all damage some of the chemical structures in tissue. If the epitope detected by a monoclonal antibody is labile in the fixative used, the antibody will not bind. A polyclonal antiserum, on the other hand, contains antibodies to various different structures and will bind to undamaged epitopes. With monoclonal antibodies it will be necessary to try different fixatives and pick the one which best preserves a particular epitope. For example, periodate-lysine-paraformaldehyde (PLP), a fixative which generally gives acceptable preservation of structure and antigenicity of proteins, destroys determinants involving sialic acid and prevents binding of antibodies recognizing epitopes which contain sialic acid. On the other hand, some monoclonal antibodies work well on standard paraffin-block material. This topic is discussed in detail in Chapter 5.

G. Labile Antibodies

A similar effect assumes importance when considering the stability of monoclonal antibodies. In a classical antiserum if some subclasses of immunoglobulin are lost relatively rapidly, a drop in titer is observed. If a monoclonal antibody happens to be an unstable protein, the loss of activity is complete.

H. Affinity

A polyclonal antibody preparation includes antibodies with a range of different affinities for the antigen. The properties of exceptionally high- or low-affinity antibodies are not noticed in such a mixture, but a monoclonal antibody of either very high or very low affinity may exhibit unusual properties and may be difficult to work with. It may be necessary to screen a number of reagents with the same specificity to select one with appropriate affinity.

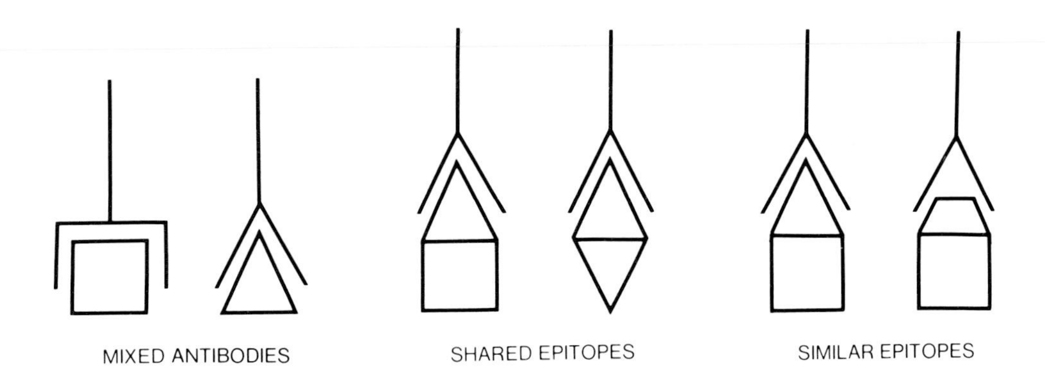


FIGURE 7. Diagrammatic illustration of the different causes of cross reactivity. Cross reactions due to mixed antibodies are absent when using monoclonal antibodies, but cross reactions due to shared and similar epitopes are found. See text for details.

### I. Cross Reactivity

A major difficulty in the use of polyclonal antisera is cross reactivity — the reaction of the serum with an antigen differing from the antigen of interest. Monoclonal antibodies are much less likely to cross react, but cross reactivity is observed and can cause incorrect interpretation of results. It is important to have a clear picture of the causes of cross reactions. These are illustrated diagrammatically in Figure 7. A common cause of cross reactivity in polyclonal antisera is the presence, in the serum, of antibodies against irrelevant antigens unrelated to the antigen of interest. This kind of cross reactivity, resulting from the presence of mixed antibodies, is absent when using monoclonal antibodies. However, a single antibody can react with unrelated tissues, cells, or macromolecules if they possess a shared epitope. Such cross reactivity is found with monoclonal as well as polyclonal antibody preparations.

An antibody may react with a structure which is similar to the epitope which it recognizes principally. Such a reaction would usually be of a lower affinity, and then whether or not cross reactivity is observed depends on the sensitivity and conditions of the assay. This kind of cross reaction is detected both by monoclonal and polyclonal antibody preparations.

Some of the limitations of monoclonal antibodies have been stressed, not to detract from the value of these reagents, but to emphasize the factors which need to be taken in consideration when assessing their suitability for use in particular situations. In spite of these limitations monoclonal antibodies are superior to conventional antisera where the two can be compared, and monoclonal antibodies can be used in situations where polyclonal antisera would not even be considered.

## V. MONOCLONAL ANTIBODIES AS STANDARD REAGENTS

Above all, monoclonal antibodies provide the opportunity for standardization. Once a clone is established, it should be possible to produce an identical antibody in unlimited amounts. This means that different laboratories can work with the same reagent, thus removing a major source of variability. The workshops on Human Leukocyte Differentiation Antigens held in Paris[2] in 1982 and Boston[3] in 1984 illustrate this aspect well. For the Paris workshop monoclonal antibodies were submitted by participating laboratories, aliquoted, and coded by the organizing laboratory, and sent out to 55 laboratories in 14 countries for analysis. Antibodies which proved to have useful specificities and to be robust, in the sense that different laboratories obtained similar results with them, can be widely adopted, since the clones are cryopreserved. A few clones have proved to be unstable, others have simply been lost (perhaps the difference is in the frankness of the originators), but on the whole *once a clone is established, an antibody will always be available without variation in specificity.*

The Paris workshop also served to illustrate one of the problems that has not been solved by monoclonal antibodies. A number of known antibodies was included in the panel, with hidden replicates. Ten percent of participating laboratories were precluded from further analysis because their performance in assessing the replicates was below the required level. Thus, *although monoclonal antibodies remove one cause of interlaboratory variation, such variation has not been totally eliminated.*

## VI. COMMENTS ON PUBLICATION, SUPPLY, AND COMMERCIALIZATION OF MONOCLONAL ANTIBODIES

Monoclonal antibodies have raised a number of questions in relation to publication of new antibodies, supply of antibody or hybridoma lines to other investigators, and commercialization. On the whole, these questions have been unexpected because of the different nature of hybridomas as compared with traditional antisera, and there has been no concensus as to answers to these questions. This section briefly discusses some of the issues.

In the first few years of hybridoma production, journals would publish any paper which described a new monoclonal antibody. Now that there are so many antibodies available, it is not enough for the authors of a paper to be saying "we also have an antibody against X." The antibody has to be against an antigen to which monoclonals have not yet been described, or the antibody must have some interesting or unexpected property. One of the problems is that it is not always easy to determine whether two antibodies are against the same antigen unless both are available together. The workshops discussed above are particularly useful in this regard, but authors should be prepared to exchange reagents to determine whether or not two antibodies differ in specificity. There are a few journals specially devoted to the publication of new hybridomas, and these provide a service in disseminating information on new hybridomas. Probably more useful in the long term will be computerized compilations of data on hybridomas useful in particular subject areas. The endocrinologist is unlikely to wish to go through data on monoclonal antibodies against human blood cells and vice versa, hence the need for specialization by subject area. Some journals require an undertaking from authors that they will make the antibody available to other workers. Such a blanket undertaking seems unreasonable and impossible to enforce; it reflects an aversion to science for profit.

This leads to the related issues of supply of material, as gifts, in exchange or commercially. There is no single "correct" approach to these issues. We have taken a flexible viewpoint, supplying antibody to many laboratories in several different countries as gifts, in exchange for other antibodies, or in collaborative work where we played an active part in the work. Because some of our reagents are potentially useful in routine diagnostic work, and because we felt that a science-based company was the best place to carry out large-scale production, packaging, quality control, and distribution, we have also had commercial agreements for marketing of our antibodies. Financial returns from such sales have been administered by the institution and used for purposes such as conference travel and purchase of journals. Academic institutions are places where profitability is regarded with great suspicion, and we have taken the view that the use of any income should be open to scrutiny in this way.

Related to the issue of commercial exploitation is the question of patentability. Kohler and Milstein published the hybridoma technique; clearly, they had no intention of taking out any patents on it. It may take lawyers a great deal of time to determine whether any subsequent patents on hybridomas are valid, but ethically they are certainly not valid, since the originators of the work did not patent it. Patents on processes or techniques which use monoclonal antibodies as reagents are a different matter, so long as the process or technique is new.

## REFERENCES

1. Kohler, G. and Milstein, C., Continuous cultures of fused cells secreting antibody of predefined specificity, *Nature (London)*, 256, 495, 1975.
2. Bernard, A., Boumsell, L., Dausset, J., Milstein, C., and Schlossman, S. F., Eds., *Leucocyte Typing*, Springer-Verlag, Berlin, 1984.
3. Reinherz, E. L., Haynes, B. F., Nadler, L. M., and Bernstein, I. D., Eds., *Leucocyte Typing II*, (3 volumes), Springer-Verlag, New York, 1986.

Chapter 2

# OVERVIEW

## I. AIM OF THE CHAPTER

In the previous chapter, the theoretical background to the use of antibodies, and particularly monoclonal antibodies, as reagents was discussed. The relative advantages of monoclonal antibodies and polyclonal antisera were considered. Subsequent chapters deal in detail with the practical aspects of making and using monoclonal antibodies.

This chapter bridges the theoretical background and the practical details. The chapter is intended to help the reader answer some questions:

- Do I need monoclonal antibodies?
- What can I realistically expect to do with them when I have them?
- Shall I make them or can I beg, borrow, or (heaven forbid) steal them?
- If I want to make my own, what do I need in the way of facilities, equipment, staff, skills, time, and maintenance budget?

The answers will obviously depend very much on the needs, interests, and inclinations of the reader. This chapter is not intended to provide the answers, but to provide the information so that the reader can make an informed decision.

## II. APPLICATIONS AND LIMITATIONS OF MONOCLONAL ANTIBODIES (DO I NEED MONOCLONALS, AND WHAT CAN I REALISTICALLY EXPECT TO DO WITH THEM?)

Monoclonal antibodies are specific probes for molecular structures such as drugs, hormones, receptors for hormones, or any of the other biologically derived or biologically active materials which interest the biological or medical scientist. A monoclonal antibody against any such structure can be used to identify it, locate it, purify it, and where appropriate, destroy it (Figure 1). Obviously, the potential uses for monoclonal antibodies are very widespread.

To translate this very broad view of the potential applications of monoclonal antibodies into a detailed analysis of their uses in each individual field would be unproductive. The reader will be expert in a particular field and will be able to make a detailed assessment for that field; in all probability the reader will not wish to go through a detailed analysis of applications pertaining to other fields. However, Table 1 contains references to reviews or key papers on the applications of monoclonal antibodies in a variety of different areas, and should enable the reader to explore what has been done in any particular field in greater detail.

The case for the potential usefulness of monoclonal antibodies in almost every field of biology and medicine is made in the above discussion and the table. Even for topics not listed, the potential uses can be appreciated by extrapolation. However, the reader will want a more critical analysis of the real difficulties associated with these potential uses, in order to make a realistic assessment of the likely benefits and costs of using this technology.

The discussion of the relative merits of monoclonal and polyclonal antibody preparations in Chapter 1 will have indicated that there are problems associated with the use of antibodies, including monoclonal antibodies, as probes for biological structures.

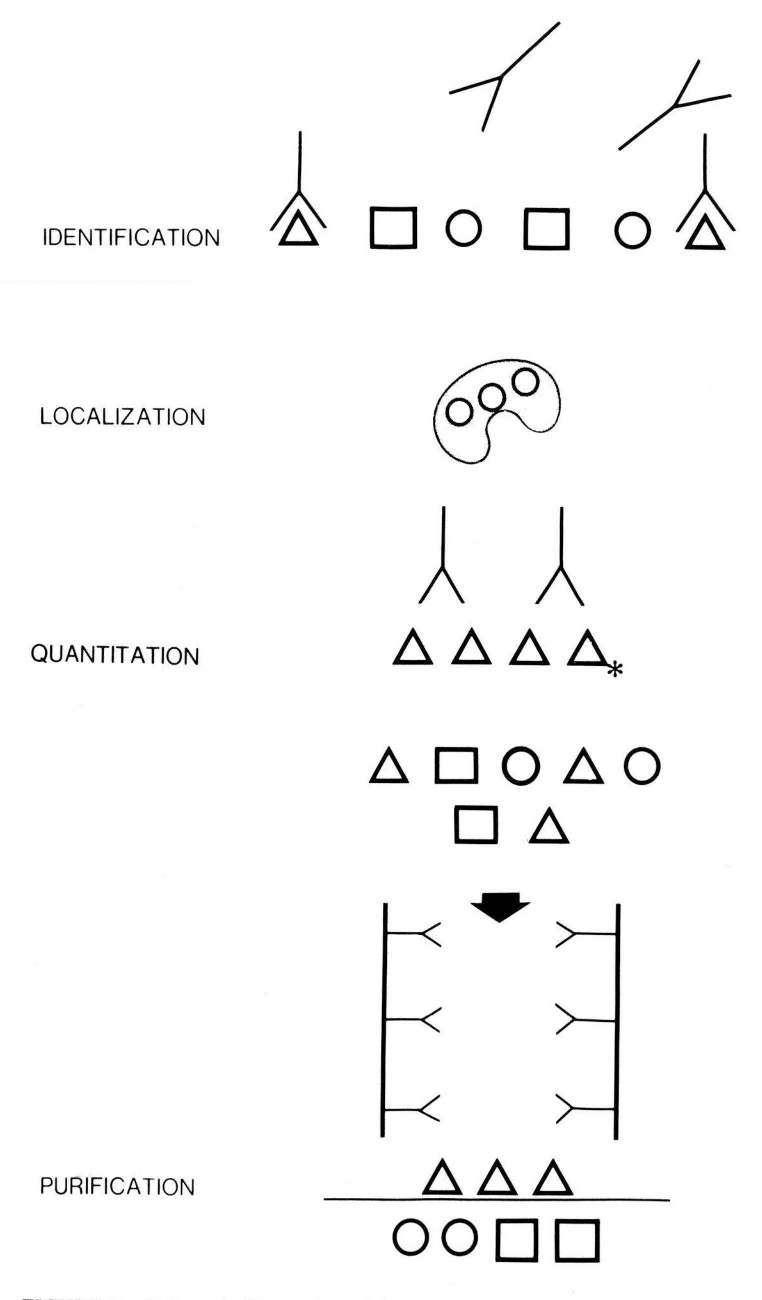


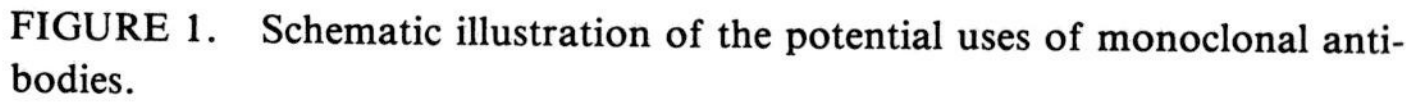

FIGURE 1. Schematic illustration of the potential uses of monoclonal antibodies.

Detailed discussion of these problems and solutions will be found in later chapters, and the aim in this section is to help the reader to decide if further exploration of this technology is likely to be rewarding. Accordingly, what follows is a summary of the potential of monoclonal antibodies in the different methodologies for which they may be considered.

## A. Identification and Quantitation of a Soluble Molecule in a Mixture

The exquisite specificity of monoclonal antibodies is a major asset when using them as reagents for identification and quantitation. This area, which is considered in detail

Table 1
REVIEWS AND KEY PAPERS ON APPLICATIONS OF MONOCLONAL ANTIBODIES

| Field or topic | Ref. |
|---|---|
| Diagnostic reagents | 1, 2 |
| Histology | 3, 4 |
| Quantitation of biologicals | Chapter 6, Table 1 |
| Purification of antigens | 5, 6 |
| Hormones | 7, 8 |
| Enzymes | 9 |
| Viruses | 10, 11 |
| Bacteria | 12 |
| Parasites | 13, 14 |
| Allergens | 15 |
| Human blood cells | 16 |
| B lymphocytes | 17 |
| T lymphocytes | 18, 19 |
| Myeloid cells | 20, 21 |
| Mouse tissue antigens | 22 |
| Rat tissue antigens | 23 |
| Cancer diagnosis | 24, 25 |
| Cancer treatment | 24, 25 |
| Leukemia/lymphoma | 26—28 |

in Chapter 6, comprises one of the major applications of monoclonal antibodies and includes the analysis of hormones, drugs, and enzymes in body fluids. One potential difficulty is that an epitope recognized by a monoclonal antibody may be found on two quite distinct molecules. In a polyclonal antiserum, antibody against such a shared epitope is less of a problem, because it is unlikely that *all* the antibodies in the serum will be directed at shared epitopes. Monoclonal antibodies are best suited to enzyme-linked immunoassay or radioisotope-based assays and, generally, are not suited to use in precipitation reactions, for reasons discussed in Chapter 1.

### B. Purification of a Soluble Molecule from a Mixture

Monoclonal antibodies are potentially very powerful reagents for preparative affinity chromatography. The potential has been demonstrated in a few instances with impressive single-step purifications of trace components from complex mixtures. Monoclonal antibody-based affinity purification, discussed in detail in Chapter 6, is the method of choice for anyone embarking on a difficult purification problem.

### C. Detection and Quantitation of a Cell Type in Suspension

A major part of the early effort with monoclonal antibodies has been concerned with the identification of cells, such as lymphocyte subpopulations. The combination of the specificity of monoclonal antibodies with the quantitative analysis possible by flow cytometry has been used extensively in immunology and hematology. Suspension studies on cells which normally occur in solid tissue are less easy, because the treatment needed to dissociate the tissue may affect the cell surface markers which are the targets for the monoclonal antibodies. In embarking on studies of cell surface markers it is important to have realistic expectations. Setting out to make reagents which will distinguish closely related cells (for example, malignant from normal) is fine, but what if the cells do not have any antigenic differences? Realistic expectations and critical evaluation of results are also needed in using monoclonal antibodies available from other projects. A monoclonal antibody may react with normal T lymphocytes and fail to react with normal B lymphocytes. If you apply it to leukemic cells, does a positive reaction mean that the leukemic cells are of the T lineage? It does not and such extrapolation is as hazardous in this field as in any other. Monoclonal antibodies have had an enormous and positive impact on cellular studies, but realistic expectations and cautious interpretation are still needed. The techniques used in this area are reviewed in Chapter 5.

### D. Separation of a Particular Cell Type from a Mixture

This is an extension of the previous section — if a cell can be distinguished from its companions using a monoclonal antibody against a surface molecule, it can, in principle, be separated from the other cells. Fluorescence-activated cell sorting has been used with great success in this area, but is limited in scale. Techniques capable of separating larger numbers of cells are attended with considerable technical problems, so that results vary widely between different laboratories. The potential of the technique is, however, great and extends to clinically useful fractionation of bone marrow cells for transplantation. This area is covered in more detail in Chapter 5.

### E. Identification of Molecular and Cellular Structures in Tissue

The requirements and difficulties are similar to those discussed for cells in suspension, with the added question of tissue fixatives. Tissue applications of monoclonal antibodies were relatively slow to get off the ground, but the development of suitable fixatives, and staining procedures which take advantage of the specificity of monoclonal antibodies (low background) and compensate for their insensitivity relative to polyclonal sera (because they react with single rather than multiple determinants) have led to the rapid acceptance of monoclonal antibodies as reagents in histology. Techniques range from optical to electron microscope levels of magnification and provide combined staining for a monoclonal antibody-detected component and other components, such as enzyme markers, as well as morphology. The finding of unexpected cross reactivities calls for caution in the interpretation of results in immunohistochemistry. For example, several markers of particular hemopoietic cell types also stain particular areas of the kidney. The identification of parasitic microorganisms in sections of human and animal tissue is a particularly promising application of monoclonal antibody techniques. The same techniques are potentially useful in the identification of plant pathogens. Immunohistochemical methods are discussed in Chapter 5.

### F. Clinical Applications

The analytical applications discussed in the preceding paragraphs clearly have potential and actual use in diagnosis. Monoclonal antibodies tagged with radioisotopes are being investigated for in vivo diagnosis, in particular, the detection of secondary tumors. The use of monoclonal antibodies for treatment of cancer, infectious diseases, and poisoning of various types is at a relatively early stage, but is the subject of much current effort. The fact that the monoclonal antibody, if genetically foreign to the patient, will itself become the target of an immune response has caused some anxiety and directed efforts to the production of human monoclonal antibody. While much of the emphasis so far has been in human medicine, the same techniques can be used in veterinary and plant pathology. These questions are considered in Chapters 5, 6, and 7.

## III. AVAILABILITY (SHALL I MAKE MY OWN MONOCLONAL ANTIBODIES OR CAN I OBTAIN THEM?)

The question of availability is best answered by scouring the commercial catalogs and the various hybridoma compilations (see Appendix 1). Since many monoclonal antibodies are neither available commercially nor listed in compilations, there is a large element of chance in finding someone with the right antibody, and the "grapevine" is an important source of information. The author's views on publication, commercial exploitation, and giving away antibodies are outlined in Chapter 1 (Section VI). In practice, if an antibody is absolutely central to the work and if more than analytical quantities are needed, the user will need to have access to the hybridoma line. This means either obtaining it from a source such as the American Type Culture Collection (Appendix 1), asking a colleague for the line, or doing some fusions.

## IV. RESOURCES NEEDED TO MAKE MONOCLONAL ANTIBODIES

This section sets out the equipment, skills, and staff needed to embark on the production of hybridomas and monoclonal antibodies. Reagents, disposable plastic ware, and other materials are listed in the later chapters. Budgeting for such materials will depend on the specific details of the project, but hybridoma work consumes large quantities of media and sterile disposable plastics. As a guide, a project which employs one technician carrying out hybridoma work will need a consumables budget approximately equivalent to the technician's salary. The reader will be able to budget much more accurately after reading the subsequent chapters and considering the detailed needs of the specific project. Appendix 2 contains a directory of suppliers of equipment and materials needed in hybridoma studies.

### A. Major Equipment

The equipment required for hybridoma work consists essentially of tissue culture apparatus and the equipment needed for antibody detection.

#### *1. Cabinet for Sterile Work*

Sterile work stations consist of a working area across which a current of sterile filtered air is maintained. The sterile air is blown over the work towards the user. This is the most satisfactory system in terms of sterility and ease of use, but if the project involves pathogenic material a biohazard cabinet should be used. In this equipment a vertical curtain of air acts as a barrier between the work and the outside world. This system prevents entry into the work area of contaminants such as spores, and exit from the work area of any pathogen. However, the system is more difficult to use effectively, and care is needed in use to avoid disturbing the laminar flow of the air curtain, for example, by moving the hand in and out too quickly. What should be considered potentially pathogenic is sometimes a matter of opinion. For example, human cell lines of tumor origin may be considered as possible carriers of retroviruses, as yet unidentified.

To maintain sterility a UV light is left on when the cabinet is not in use, but must be switched off during use, since the UV light is damaging to the eyes and skin. A plastic UV-absorbing curtain is used when the UV light is on to protect staff working nearby. The work area should be kept clean by washing down with alcohol before or after each use, and the filtered air should be switched on 30 min before work starts. The cabinets should be properly maintained according to the manufacturer's instructions to ensure that they are effective.

*2. Gassed Incubator*

The media used for most tissue culture work depend on carbonate/bicarbonate buffering and require a $CO_2$-enriched atmosphere. In the early stages of hybridoma cultures they are best kept in unsealed culture vessels, in equilibrium with the humidified, $CO_2$-enriched atmosphere. Thus, the incubator used must control temperature, humidity, and $CO_2$ concentration. Many incubators are available, but it is important to use one of good quality. Points to look for are (1) temperature control should be precise (if the temperature creeps up to 41°C, cultures can be lost); (2) the $CO_2$ concentration should be readjusted rapidly after the doors have been opened; and (3) it is advantageous to use an incubator with two separate sections so that cultures which are particularly sensitive to pH fluctuation can be kept separately from cultures which require frequent access. Corrosion has been a problem in some makes of incubator sold in recent years.

*3. Liquid Nitrogen Facility*

Hybridomas are stored at liquid nitrogen temperatures. A large vacuum-jacketed liquid-nitrogen tank, preferably with a convenient system of racks or drawers is needed together with a supply of liquid nitrogen. A programed cooling device is useful but not essential (see Chapter 3, Section X).

*4. Animal House*

Access to animal holding facilities will be required. Only small numbers of animals are needed, usually mice, for immunization, as a source of feeder cells, and for the production of ascites fluid.

B. Minor Equipment

The following list concerns items needed especially for hybridoma work. The laboratory will need general equipment, such as balances, a pH meter, and a supply of distilled water, as well as washing-up facilities.

1. Inverted microscope with phase optics to examine cultures.
2. Laboratory microscope and hemocytometers to count cells and determine viability.
3. Laboratory centrifuge, preferably refrigerated.
4. Water bath kept at 37°C.
5. Sterile surgical equipment for mouse dissection.
6. Equipment to sterilize medium; disposable sterile filters can be used for small volumes; most laboratories will also find it economical to make up (from ready-mixed powders) and sterilize their own media, and this requires larger-scale sterilization equipment.
7. Facilities for sterilizing pipettes; alternatively, sterile disposable pipettes may be used, at greater expense.

In addition to the equipment listed, other items, large and small, will be needed for screening of antibody activity. The nature of this equipment depends on the assays chosen (see Chapter 3, Section VIII). Additional equipment will be needed to characterize and purify antibody (Chapter 4) and to use it (Chapters 5 and 6).

C. Staff and Skills Required

Hybridoma work requires a certain understanding of immunology in order to pick the most appropriate solutions to problems. However, an extensive knowledge of immunology is not required, and this book attempts to provide the necessary background

for the nonimmunologist. What is more important is that at least one member of a group undertaking a hybridoma project should have some experience in cell culture. A number of books covering cell culture methods, in general, are available (see Further Reading at the end of this chapter) and cell culture techniques may be learned by spending a few weeks in another laboratory. Alternatively, it may be possible to employ a technician with appropriate experience, or to send a technician to a hybridoma group for training. If the group setting up the project has no experience in this area at all, a period from 1 to 3 months would be suitable.

Thus, the hybridoma techniques themselves can be learned by any competent biological scientist. However, it is essential to be expert with the antigen that is the subject of the project. While specific examples of the production of monoclonal antibodies against different types of antigen will be discussed later in this book, the production of the antigen and the choice of the most appropriate assays will depend on expertise with the antigen. For this reason it is generally not advisable for a "hybridoma expert" to scan around for antigens against which monoclonal antibodies might be made. Nor is it productive for a scientist who has a need for monoclonal antibodies to ask a hybridoma laboratory to take on the entire project. The best approach is for the group who has the need for monoclonal antibodies to either set up the method in their laboratory (after reading this book) or to arrange an active collaboration with a hybridoma laboratory, where antigen preparation and antibody screening are carried out by the laboratory with expertise with the antigen.

The staff required to carry out a hybridoma project needs to have or to acquire cell culture experience. A careful, well-organized individual can carry out the work without having any detailed knowledge either of immunology or of the antigen, provided the important decisions are taken by people who do have such knowledge. One person can carry out the production and characterization of monoclonal antibodies, provided the volume of work is kept within bounds. The nature of the work, as will be apparent in subsequent chapters, leads to frantically busy periods when hybridomas are being screened, cloned, and cryopreserved, while at other times, when waiting for animals to be immunized or hybrids to start growing, there is time to fit in other work.

## REFERENCES


1. McMichael, A. J. and Fabre, J. W., Monoclonal antibodies, in *Clinical Medicine,* Academic Press, London, 1982.
2. Zola, H., Speaking personally: monoclonal antibodies as diagnostic reagents, *Pathology,* 17, 53, 1985.
3. Warnke, R. and Levy, R., Detection of T and B cell antigens with hybridoma monoclonal antibodies: a biotin-avidin-horseradish peroxidase method, *J. Histochem. Cytochem.,* 28, 771, 1980.
4. Poppema, S., Bhan, A. K., Reinherz, E. L., McCluskey, R. T., and Schlossman, S. F., Distribution of T cell subsets in human lymph nodes, *J. Exp. Med.,* 153, 30, 1981.
5. Parham, P., Monoclonal antibodies against HLA products and their use in immunoaffinity purification, in *Methods in Enzymology,* Vol. 92, Langone, J. J. and Van Vunakis, H., Eds., Academic Press, Orlando, 1983, 110.
6. Goding, J. W., *Monoclonal Antibodies: Principles and Practice,* Academic Press, New York, 1983.

7a. Ivanyi, J., Analysis of monoclonal antibodies to human growth hormone and related proteins, in *Monoclonal Hybridoma Antibodies: Techniques and Applications,* Hurrell, J. G. R., Ed., CRC Press, Boca Raton, Fla., 1981, 59.

7b. Kupchik, H. Z., Antibodies to alphafetoprotein and carcinoembryonic antigen produced by somatic cell fusion, in *Monoclonal Hybridoma Antibodies: Techniques and Applications,* Hurrell, J. G. R., Ed., CRC Press, Boca Raton, Fla., 1981, 81.

8. Fellows, R. E. and Eisenbarth, G. S., *Monoclonal Antibodies in Endocrine Research,* Raven Press, New York, 1981.
9. Harris, H., Monoclonal antibodies to enzymes, in *Monoclonal Antibodies and Functional Cell Lines Progress and Applications,* Kennett, R. H., Bechtol, K. B., and McKearn, T. J., Eds., Plenum Press, New York, 1984, 33.
10a. Laver, W. G., The use of monoclonal antibodies to investigate antigenic drift in influenza virus, in *Monoclonal Hybridoma Antibodies: Techniques and Applications,* Hurrell, J. G. R., Ed., CRC Press, Boca Raton, Fla., 1981, 103.
10b. Pereira, L., Monoclonal antibodies to herpes simplex viruses 1 and 2, in *Monoclonal Hybridoma Antibodies: Techniques and Applications,* Hurrell, J. G. R., Ed., CRC Press, Boca Raton, Fla., 1981, 119.
11a. Nowinski, R. C., Stone, M. R., Tam, M. R., Lostrom, M. E., Burnette, W. N., and O'Donnell, P. V., Mapping of viral proteins with monoclonal antibodies. Analysis of the envelope proteins of murine leukemia viruses, in *Monoclonal Antibodies. Hybridomas: A New Dimension in Biological Analyses,* Kennett, R. H., McKearn, T. J., and Bechtol, K. B., Eds., Plenum Press, New York, 1980, 295.
11b. Gerhard, W., Yewdell, J., Frankel, M. E., Lopes, A. D., and Staudt, L., Monoclonal antibodies against influenza virus, in *Monoclonal Antibodies. Hybridomas: A New Dimension in Biological Analyses,* Kennett, R. H., McKearn, T. J., and Bechtol, K. B., Eds., Plenum Press, New York, 1980, 317.
11c. Koprowski, H. and Wiktor, T., Monoclonal antibodies against rabies virus, in *Monoclonal Antibodies. Hybridomas: A New Dimension in Biological Analyses,* Kennett, R. H., McKearn, T. J., and Bechtol, K. B., Eds., Plenum Press, New York, 1980, 335.
12. Polin, R. A., Monoclonal antibodies against streptococcal antigens, in *Monoclonal Antibodies Hybridomas: A New Dimension in Biological Analyses,* Kennett, R. H., McKearn, T. J., and Bechtol, K. B., Eds., Plenum Press, New York, 1981, 353.
13. Mitchell, G. F., Hybridomas in immunoparasitology, in *Monoclonal Antibodies and Functional Cell Lines: Progress and Applications,* Kennett, R. H., Bechtol, K. B., and McKearn, T. J., Eds., Plenum Press, New York, 1984, 139.
14. Phillips, S. M. and Zodda, D. M., Monoclonal antibodies and immunoparasitology, in *Monoclonal Antibodies and Functional Cell Lines: Progress and Applications,* Kennett, R. H., Bechtol, K. B., and McKearn, T. J., Eds., Plenum Press, New York, 1984, 239.
15. Smart, I. J., Heddle, R. J., Zola, H., and Bradley, J., Development of monoclonal mouse antibodies specific for allergenic components in ryegrass (Lolium perenne) pollen, *Int. Arch. Allergy Appl. Immunol.,* 72, 243, 1983.
16. Bernard, A., Boumsell, L., Dausset, J., Milstein, C., and Schlossman, S. F., *Leucocyte Typing. Human Leucocyte Differentiation Antigens Detected by Monoclonal Antibodies,* Springer-Verlag, Berlin, 1984.
17. McKenzie, I. F. C. and Zola, H., Monoclonal antibodies to B cells, *Immunol. Today,* 31, 10, 1983.
18. Reinherz, E. L., Kung, P. C., Goldstein, G., Levey, R. H., and Schlossman, S. F., Discrete stages of human intrathymic differentiation: analysis of normal thymocytes and leukemic lymphoblasts of T-cell lineage, *Proc. Natl. Acad. Sci. U.S.A.,* 77, 1588, 1980.
19. Acuto, O. and Reinherz, E. L., The human T-cell receptor. Structure and function, *N. Engl. J. Med.,* 312, 1101. 1985.
20. Andrews, R. G., Brentnall, T. A., Torok-Storb, B., and Bernstein, I. D., Stages of myeloid differentiation identified by monoclonal antibodies in leucocyte typing, in *Human Leucocyte Differentiation Antigens Detected by Monoclonal Antibodies,* Bernard, A., Boumsell, L., Dausset, J., Milstein, C., and Schlossman, S. F., Eds., Springer-Verlag, Berlin, 1984, 398.
21. Polli, N., Zola, H., and Catovsky, D., Characterization by ultrastructural cytochemistry of normal and leukemic myeloid cells reacting with monoclonal antibodies, *Am. J. Clin. Pathol.,* 82, 389, 1984.
22a. Ledbetter, J. A., Goding, J. W., Tokuhisa, T., and Herzenberg, L. A., Murine T-cell differentiation antigens detected by monoclonal antibodies, in *Monoclonal Antibodies. Hybridomas: A New Dimension in Biological Analyses,* Kennett, R. H., McKearn, T. J., and Bechtol, K. B., Eds., Plenum Press, New York, 1981, 235.
22b. Springer, T. A., Cell surface differentiation in the mouse. Characterization of "jumping" and "lineage" antigens using xenogeneic rat monoclonal antibodies, in *Monoclonal Antibodies. Hybridomas: A New Dimension in Biological Analyses,* Kennett, R. H., McKearn, T. J., and Bechtol, K. B., Eds., Plenum Press, New York, 1981, 185.
23. Mason, D. W., Brideau, R. J., McMaster, W. R., Webb, M., White, R. A. H., and Williams, A. F., Monoclonal antibodies that define T-lymphocyte subsets in the rat, in *Monoclonal antibodies. Hybridomas: A New Dimension in Biological Analyses,* Kennett, R. H., McKearn, T. J., and Bechtol, K. B., Eds., Plenum Press, New York, 1981, 251.

24. Mitchell, M. S. and Oettgen, H. F., *Progress in Cancer Research and Therapy. Hybridomas in Cancer Diagnosis and Treatment,* Vol. 21, Raven Press, New York, 1982.
25. Boss, B. D., Langman, R., Trowbridge, I., and Dulbecco, R., *Monoclonal Antibodies and Cancer,* Academic Press, Orlando, 1983.
26. Zola, H. and Kupa, A., The use of monoclonal antibodies in the analysis of leukemia and lymphoma, *Dis. Markers,* 1, 117, 1983.
27. Greaves, M. F., Delia, D., Robinson, J., Sutherland, R., and Newman, R., Exploitation of monoclonal antibodies: a 'Who's Who' of haemopoietic malignancy, *Blood Cells,* 7, 257, 1981.
28. Nadler, L. M., Reinherz, E. L., Weinstein, H. J., D'Orsi, C. J., and Schlossman, S. F., Heterogeneity of T cell lymphoblastic malignancies, *Blood,* 55, 806, 1980.

## FURTHER READING

Basic Immunology

Roitt, I., *Essential Immunology,* 5th ed., Blackwell Scientific, Oxford, 1984.

Cell Culture

Adams, R. L. P., *Cell Culture for Biochemists, Laboratory Techniques in Biochemistry and Molecular Biology,* Burdon, R. and Von Kippenberg, P. H., Eds., Elsevier/North-Holland, Amsterdam, 1980.

Freshney, R. I., *Culture of Animal Cells. A Manual of Basic Technique,* Alan R. Liss, New York, 1983.

Chapter 3

# MAKING HYBRIDOMAS

## I. SCOPE OF THE CHAPTER

You have decided that you need to make your own monoclonal antibodies. This chapter assumes that you have the antigen (either pure or as a component of a mixture) and deals with immunization, hybridization, screening, cloning, and preservation of the clones. The chapter provides a detailed practical guide through the technical obstacle course that stands between possession of the antigen and the establishment of a secure bank of cloned hybridoma cells which make antibody against the antigen.

There are several alternative ways to carry out each particular step in the process. This chapter will set out, in detailed recipe form, methods which the author knows work from personal experience. Alternatives will be discussed, and sufficient information will be provided to enable the reader to make rational choices between alternatives. Some of the choices will depend on the antigen in question. Sufficient information will be provided to enable the reader to modify methodology to suit individual needs. This chapter deals specifically with mouse hybridomas, but references to techniques used for fusions using cells of other species are provided.

## II. STRATEGY

Making a hybridoma requires the following sequence of steps:

1. Prepare antigen (enough for immunization and screening).
2. Plan screening test. You will need to detect antibody against the antigen. If the antigen is not pure you will need a specificity control screen, which tells you when the antibody detected reacts with another component of the mixture, rather than with the antigen itself.
3. Develop the screening assay. The assay must be available in working order before fusions are initiated. A known positive and a known negative antibody are needed to develop the assay. If this is part of a continuing program an existing assay may be adapted, with reasonable confidence that it will work. If the project is designed to produce a monoclonal antibody where a polyclonal antiserum has been in use, an assay will be available, though it may need to be modified, and the polyclonal antiserum may be used as a positive control. It may be possible to obtain a small amount of monoclonal antibody from someone else to set up the assay. If none of these options are available immunization of the animals to be used for fusion should proceed, and sample bleeds taken from these animals should be used as positive controls. If the test does not work, it is not clear whether the assay or the sample bleed is at fault.
4. Immunize the animals. Most immunization protocols require a period of 4 to 5 weeks. This can be started when the development of the screening test is under way and does not look like taking longer than the immunization.
5. Carry out the first fusion. It is best to use two mice and keep one in reserve to fuse again shortly if the first hybridization fails.
6. Wait and watch, occasionally feeding the hybrids. If the fusion is contaminated this will quickly be apparent. If after 10 to 14 days there are no colonies, they may still come up, but it is time to boost the reserve mouse.

7. When colonies appear and start to yellow the medium, arrange to have few other commitments, because the next 2 weeks are likely to be busy. Colonies must be screened. Wells with colonies but a negative screening result should be allowed to overgrow slightly and be retested so as to avoid discarding them prematurely. Positive colonies can either be cloned immediately or tested further for specificity. This depends on the objectives; if they are very narrowly defined there is no point working on cross-reactive antibodies. However, a supernatant may be nonspecific because it contains two colonies. The probability that this is so can be estimated from the frequency of colonies: if there are very few, the chances of having two in the same well are small; if most wells are positive, the chances that some will contain two colonies are high (see discussion on Poisson distribution in Section IX.B.1.a.).
8. While the clones are growing up (this also takes about 10 days) it is a good idea to scale up the original wells and cryopreserve cells. If this is not done, at least the cells should be maintained in culture until it is clear that the cloning has worked.
9. Clones are screened and a few positives (two to four from each original well) are scaled up and cryopreserved.
10. Now you can catch up with other things. Given an existing screening test and antigen already available, the process as a whole takes about 5 weeks for immunization (this can be much shorter or somewhat longer depending on the protocol chosen) and 6 to 8 weeks from fusion to having the clones in the (cell) bank.

Summarizing the sequence of events:

1. Prepare antigen.
2. Plan screening assay.
3. Develop screening assay.
4. Immunize animals.
5. Hybridize.
6. Nurture hybrids.
7. Screen supernatants.
8. Clone, cryopreserve uncloned cells.
9. Screen clones. Cryopreserve positives.

Figure 1 illustrates the component parts of a hybridization experiment.

## III. MATERIALS TO BE ORDERED

### A. Quantities Needed

The quantities given below represent approximate amounts needed per fusion (of one spleen), including cloning, scale-up, and cryopreservation. These estimates are provided to allow a rough estimate of usage over the year, based on the number of fusions planned. Medium, serum, and hypoxanthine and thymidine will also be used to produce monoclonal antibody as tissue culture supernatant from established hybridomas.

1. Medium: cell preparation and fusion, 500 mℓ; feeding to screen stage, 200 mℓ; cloning, 200 mℓ; cryopreservation, 200 mℓ. Total: 1100 mℓ.
2. Fetal calf serum: 10% of total medium used, i.e., 110 mℓ.
3. HAT constituents: aminopterin, enough for 100 mℓ/fusion, i.e., 0.19 mg/fusion;

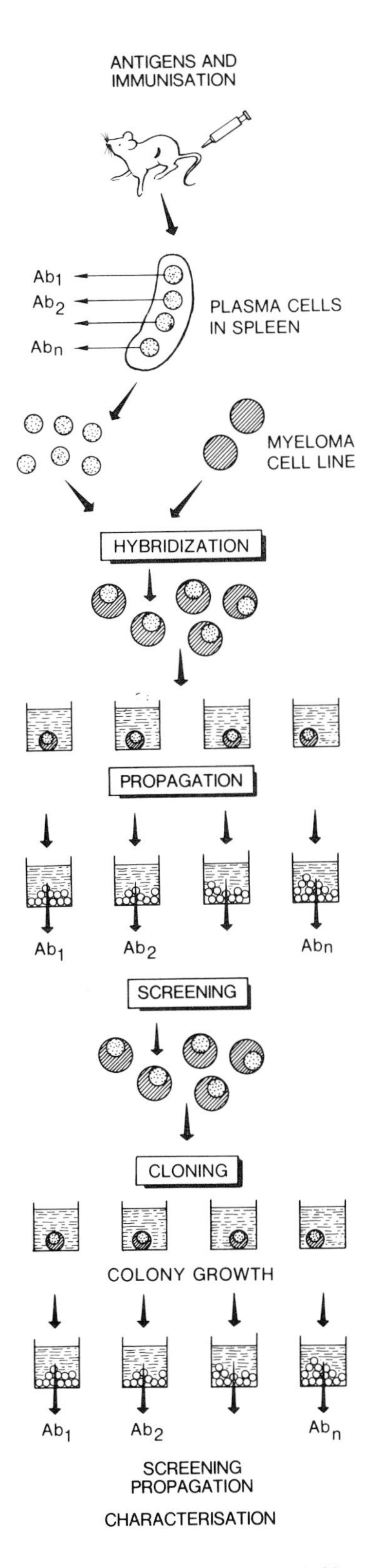


FIGURE 1. Schematic representation of hybridoma production.

hypoxanthine and thymidine (H & T), enough for 600 mℓ/fusion, i.e., 8.16 mg H and 2.4 mg T.
4. Dimethyl sulfoxide: 20 mℓ/fusion (mostly for cryopreservation).
5. Polyethylene glycol: 0.6 g/fusion.
6. Geys hemolytic medium: sufficient stock solutions (see below) for 20 mℓ/fusion (which allows three spleens for feeder cells in cloning).
7. PBS and other buffers: requirements depend on the assay selected.
8. Mice: two to three for immunization, three per fusion for feeder layers used in cloning, and more later for preparation of ascites; 25 mice per month will service an average hybridoma project.
9. Plastic disposable tissue-culture ware. Estimates of the quantity needed can at best be wild approximations. For an "average" fusion, two 24-well cluster dishes, 20 96-well cloning plates, 20 small (25 $cm^2$) tissue culture flasks, and 20 medium (80 $cm^2$) culture flasks. Freezing vials, pipettes (glass sterilizable or plastic disposable) will also be needed in large numbers. The tissue culture "novice" should obtain catalogs from the manufacturers (see Appendix 2) to see what is available and, preferably, spend some time in a hybridoma laboratory to see which plastic ware is most useful.

B. Description

*1. Cell Culture Medium: RPMI 1640 or Equivalent (See Section IV)*

These media can be purchased either in liquid form or, more economically, as powders ready to dissolve and sterilize. If the medium is to be bought as powder, a sterile filtration system of suitable capacity will be needed (see Section IV.1).

A supply of suitable water is needed if the medium is to be made up from powder. Water quality is important, and the problems caused by impure water are insiduous rather than obvious, being manifest as slow growth and low yields of hybrids. As a minimum, double-distilled water should be used. It should be used as soon as possible after distillation to avoid the growth of bacteria and consequent production of pyrogens. The use of "type I" reagent grade water is recommended. This can be prepared using water purifiers (for instance, Gelman's Water-1 System) which process the water through a cartridge which removes ions on a mixed-bed ion exchanger, organic molecules using activated charcoal, and particulate matter using a membrane filter. This final membrane filter should be an ultrafilter, with a molecular weight cutoff of 10,000 daltons, to remove pyrogen.

Glutamine and antibiotics have to be added separately and are available as sterile solutions ready to add. These materials should be kept frozen, since they lose activity at 4°C.

*2. Fetal Calf Serum (FCS) or Other Serum*

Serum is used to support growth of cells in culture. Different batches of fetal calf serum support growth of cells to a different degree, and a batch that supports growth of one cell type will not necessarily work with another cell type. Thus, it is necessary to screen serum batches before selecting one for purchase, or to buy material already selected for hybridoma use.

It is a good idea to obtain some serum from a laboratory that is already getting good yields of hybrids for the early fusions in a new project, to eliminate this variable from the list of possible causes of failure.

The most rigorous way to screen serum is to try several (four to six) batches and to distribute the products from a single fusion into plates in medium containing the different sera. The first few days after fusion are critical, and this is the most discriminating test for serum. However, in practice most laboratories screen sera by testing their ability to support growth of either an established hybridoma line or the myeloma line being used in the fusion experiments.

Mixtures of fetal calf serum and horse serum may be used, and there is a number of serum-free media available for hybridoma culture (see Chapter 4, Section II.A.2).

*3. HAT Selective Medium Components*

Hypoxanthine, aminopterin, and thymidine are used to obtain selective growth of hybrids (see Section VI.A).

*4. Dimethyl Sulfoxide*

Dimethyl sulfoxide is used in the fusion mixture and also in cryopreservation of hybridomas.

*5. Polyethylene Glycol (PEG)*

This material is the crucial fusion-inducing agent. It is very variable and it is advisable to start with some material from a colleague who is already using it for hybridization. PEG is toxic to cells, and both its toxicity and its ability to promote fusion vary from batch to batch. One important variable is the molecular weight, and the figure given on the label is only an approximate indication of an average. Molecular weights from 600 to 6000 are commonly used for hybridization.

## IV. STOCK AND WORKING SOLUTIONS

Note that weights quoted in the recipes are for anhydrous salts, unless the water of crystallization is stated.

### A. Medium

RPMI 1640 medium is either purchased in sterile liquid form, or made up from powder, sterilized by membrane filtration and dispensed into 500-mℓ bottles. Sterile filtration equipment is available from a number of manufacturers (Millipore, Gelman, Sartorius, and others) covering a range of volumes. The equipment supplier's instructions should be followed carefully, especially in testing the integrity of the filter after filtration is complete. This test depends on the fact that membrane filters of the pore size used for sterilization do not pass air when wet unless a threshold pressure (the bubble point) is exceeded. Thus, if the membrane passes air after the liquid has been filtered and before the pressure has been raised to the bubble point, the membrane has been perforated.

A thorough quality-control program should be instituted to check the sterility and other essential properties of each new batch of medium (see Chapter 4, Section II.A).

The medium may be stored in sealed bottles in a 4°C room or refrigerator indefinitely. Before use the following ingredients, which do not have a long shelf life at 4°C, are added: glutamine to 2 m*M*, penicillin 100 IU/mℓ, streptomycin 100 µg/mℓ, FCS to 10%. For convenience, the penicillin, streptomycin, glutamine supplement is referred to as PSG and the supplemented medium (medium + PSG + FCS) is referred to as complete medium. Alternative antibiotics are discussed in Chapter 4, Section II.A.3.c). complete medium. Alternative antibiotics are discussed in Chapter 4, Section I.A.4.c). If the medium is then stored for 2 weeks or more the level of PSG should be replenished, because these ingredients have a short life. A bottle of medium in use may be kept in the refrigerator, but as the liquid level falls and the volume of air in the bottle increases the medium will become increasingly alkaline (the indicator color will change from orange to red), and such medium should be adjusted by gassing with sterile carbon dioxide or by leaving it in the gassed incubator, with the bottle top loose, before use. It is safer to use fresh medium for fusion or cloning, although a well-established culture of rapidly growing cells will tolerate slightly alkaline medium. Medium should be stored in the dark, since light initiates the formation of substances which are toxic to cells. Refrigerated medium should be warmed to 37°C in a water bath before being added to cultures.

There seem to be no consistent advantages to using any of the alternative media, or to using sera other than FCS. However, cells accustomed to one medium may take some time to adjust to a different medium and should be "weaned" on to a new medium gradually by removing 50% of the old medium and replacing with new medium, and repeating this process every time the culture is ready to be subdivided.

### B. HAT Medium

Prepare a 100× stock solution consisting of 1.36 mg hypoxanthine, 0.019 mg aminopterin, and 0.388 mg thymidine/mℓ water. If the aminopterin does not dissolve, make the pH slightly alkaline by addition of 1 *N* NaOH (dropwise, with mixing). Sterilize and store frozen. For use, add 1 mℓ stock solution per 100 mℓ medium.

### C. HT Medium

Prepare a stock as HAT but with aminopterin omitted, sterilize, and store frozen. For use, add 1 mℓ/100 mℓ complete medium.

### D. Geys Hemolytic Medium

This is a buffer used to lyse erythrocytes in the spleen cell suspension. This buffer may cause less damage to lymphoid cells than alternative hemolytic formulations, such as isotonic ammonium chloride or distilled water. Geys hemolytic medium should not be confused with Geys growth medium. Stock solutions are made up as follows.

#### *1. Geys Solution A*

| | |
|---|---|
| Ammonium chloride $NH_4Cl$ | 35.0 g |
| Potassium chloride KCl | 1.85 g |
| Disodium hydrogen orthophosphate $Na_2HPO_4 \cdot 12H_2O$ | 1.50 g |
| Potassium dihydrogen orthophosphate $KH_2PO_4$ | 0.119 g |
| Glucose | 5.0 g |
| Phenol red | 0.05 g |
| Gelatine (Difco) | 25.0 g |
| Distilled water | 1000 mℓ |

Gelatine is a variable product. Although many different preparations will probably work equally well, we have used the Difco preparation routinely.

The ingredients are dissolved, in turn, in the water and the mixture is dispensed into 20-mℓ aliquots in autoclavable vials and autoclaved at 15 psi for 15 min. It can then be stored at room temperature indefinitely.

#### *2. Geys Solution B*

| | |
|---|---|
| Magnesium chloride $MgCl_2 \cdot 6H_2O$ | 4.20 g |
| Magnesium sulfate $MgSO_4 \cdot 7H_2O$ | 1.40 g |
| Calcium chloride $CaCl_2$ | 3.40 g |
| Distilled water | 100 mℓ |

Dissolve salts with stirring, dispense into 10-mℓ autoclavable vials, and autoclave at 10 psi for 10 min. This solution keeps indefinitely at room temperature.

*3. Geys Solution C*

Sodium bicarbonate solution 5.6% is Gey's solution C. This is available from media suppliers or can be made up as follows:

| | |
|---|---|
| Sodium bicarbonate $NaHCO_3$ | 5.6 g |
| Distilled water | 100 mℓ |

This solution keeps if tightly stoppered without a large air space, but otherwise will gradually lose carbon dioxide and become more alkaline.

*4. Working Solution*

Make up no more than 30 min before use:

| | |
|---|---|
| Distilled water | 14.5 mℓ |
| Gey's solution A | 4 mℓ |
| Gey's solution B | 1 mℓ |

Gey's solution C added dropwise to adjust pH to 7.2—7.4 (judged by color — should be about the same as the RPMI medium — if uncertain check with a pH meter until accustomed to the indicator color)

The amount of solution C to be added depends on how old this solution is, but should be no more than 0.5 mℓ.

The working solution should not need sterilization, since the ingredients are all sterile. If in doubt sterilize through a membrane.

E. Phosphate-Buffered Saline (PBS)

Dulbecco's PBS is used free of Ca and Mg, since these cations promote cell clumping. It may be bought from media manufacturers or made up as follows:

| | |
|---|---|
| NaCl | 8.0 g |
| KCl | 0.2 g |
| $Na_2HPO_4$ | 1.15 g |
| $KH_2PO_4$ | 0.2 g |

Dissolve in sequence in 800 mℓ distilled water. Adjust pH to 7.3 with 1 mℓ HCl or NaOH and make volume up to 1000 mℓ with distilled water. Sterilize by autoclaving at 15 psi for 15 min. It is convenient to make 10- to 20-ℓ batches of this buffer.

The pH and osmolarity should be checked. Most large biochemistry departments have instruments which measure freezing-point depression rather than osmotic pressure, but give a reliable measure of osmolarity. Variation in osmolarity of PBS is a frequent source of problems with cell viability.

F. PEG/DMSO Fusing Agent Solution

Weigh 10 g PEG into a glass screw-cap bottle. Autoclave (15 psi/15 min) to liquefy and, while the material is still warm and liquid, add 14 mℓ of a solution made by adding 3 mℓ DMSO to 17 mℓ sterile PBS. Mix well and store at 4°C. The final composition of this solution is 42% PEG by weight and 15% DMSO by volume. Check the pH of the solution. Acid fusing solution will damage the cells, and we use a pH range between 7.2 and 8.0. Westerwoudt[1] recommends an alkaline pH, around 8.2.

## V. IMMUNIZATION

Immunization comprises the first stage of a hybridoma production project (Figure 1). Monoclonal antibodies can be made against a wide range of different substances,

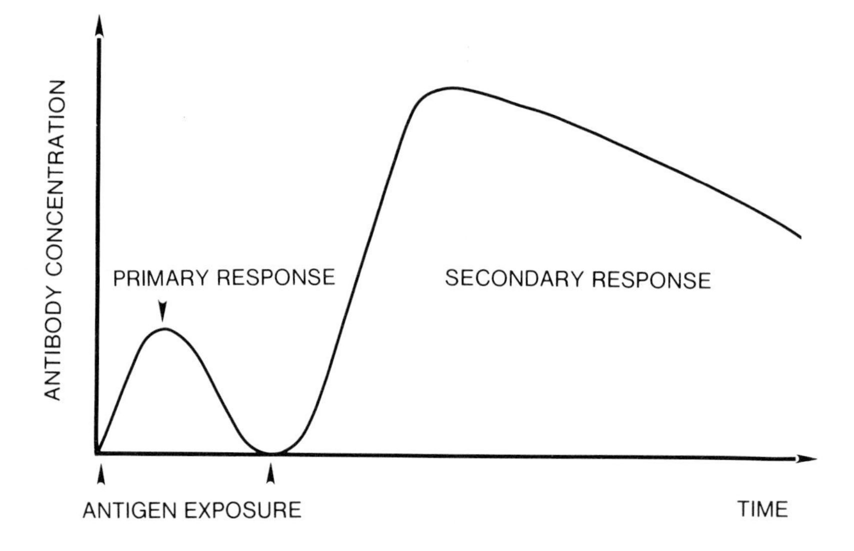


FIGURE 2. Kinetics of antibody responses: initial exposure to antigen results in a short-lived, low-grade IgM antibody response. Subsequent exposure to the same antigen results in a long-lived response, and the IgG antibody is produced in greater quantity and has a higher affinity for the antigen.

and the immunization schemes which can be used are correspondingly diverse. There is no single immunization protocol which is universally optimal. There are, however, some general principles which may be used to select an immunization protocol which is suited to the antigen in question. These principles will be discussed in this section.

### A. The Antibody Response

An animal can make antibody against a very wide range of molecular structures, including protein, carbohydrate, lipid, and nucleic acid. Small molecules, such as amino acids, monosaccharides, and synthetic compounds such as drugs, can elicit an antibody response, although they may need to be conjugated to a protein "carrier" in order to do so. In general, an animal will not make antibody against components of its own tissue. There are important exceptions to this rule and ways of circumventing it, but, by and large, if we wish to make an antibody against a tissue or serum antigen we must immunize an animal which lacks that antigen.

The immune response is adaptive. If an animal is injected with an antigen which it has never encountered before, it will make a small amount of antibody (primary response). If, some time later, the same antigen is reinjected, the antibody response (secondary response) is faster, stronger (more antibody is made), and qualitatively different (different types of antibody molecules are made and they generally bind the antigen with a higher affinity, as compared with the primary response). The kinetics of an antibody response are illustrated in Figure 2. Immunization protocols should take these features of the antibody response into consideration.

### B. A "Typical" Immunization Protocol

Given a new antigen and no information on its properties, the immunization protocol would be something like this:

1. Prime mouse.
2. Boost 4 weeks later.
3. Sacrifice mouse 4 to 5 days after boost, taking spleen for fusion and blood to test for serum antibody.

C. List of Variables

The variables in this protocol may be listed as follows:

1. Choice of species and strain of animal.
2. Purity and dose of antigen, route of administration, use of adjuvant, and attempts to make response specific to a particular antigen in a mixture.
3. Length of time interval, number of booster injections, dose and route of boosters, use of adjuvant, serial bleedings to monitor antibody response.
4. Time interval after final boost, selection of animal from a group on the basis of serum antibody titer.

D. Detailed Analysis of Variables

*1. Choice of Animal*

Most work to date has been done with mice. The rat is useful when making antibody against mouse antigens, and it is also said to be advantageous because the size of the animal allows the production of larger volumes of ascites. For the present, however, so much more is known on the use of mice and mouse myelomas that if there is no particular reason for choosing another species the mouse should be used. Human monoclonal antibodies have particular advantages in some situations, which will be discussed in Chapter 7. Practical details for the preparation of human hybridomas may be found in Olsson and Kaplan[2] and Buck et al.[3]

The myelomas in general use are from the BALB/c strain of mice, and it is thus convenient to use this strain for immunization. Other strains are used for particular purposes, for example, in the production of antibody against genetically polymorphic mouse proteins, or immunization with antigens which are poorly immunogenic in BALB/c mice. In particular, the Biozzi high-response strain of mouse has been used extensively without any convincing evidence that hybridoma yields are improved.

*2. Antigen*

If the antigen is available in pure form it should be used pure, to reduce the number of irrelevant hybrids produced. Antigenic competition is a phenomenon which is poorly understood, but in some situations the immune response to the antigen of interest may be suppressed by the presence of other antigens in the immunizing mixture. This phenomenon is by no means universal, as demonstrated by the successful routine immunization of children with multiple bacterial and viral antigens administered simultaneously.

One of the major advantages of the hybridoma technique over classical antibody production is that the antigen does not need to be pure for immunization, so long as it is possible to design a screening assay that distinguishes antibody against the antigen of interest from antibody against contaminants.

*3. Dose, Adjuvant, Route*

If the antigen is a cell surface antigen and the whole cell is being used as immunogen, a dose of from 1 to 20 million cells may be used, the higher numbers being preferred if the cells are not in short supply. If the cells are alive and proliferate in the host for a short time the immune response is more intense. Adjuvants may be used with cellular antigens, but are generally not necessary. Soluble protein antigens may be injected at doses from 5 μg to 5 mg, depending on availability, immunogenicity, and purity. Clearly, if the antigen of interest comprises only 10% of the protein in the immunizing mixture, the total weight required to give the animal a particular dose of the antigen will be ten times higher than if the antigen is pure. Adjuvants are usually helpful with soluble antigens, particularly when the amount of antigen given is small. A variety may be used, but none approaches the efficacy of Freund's complete adjuvant (FCA). Immunization may be by the intraperitoneal (i.p.), intravenous (i.v.), subcutaneous (s.c.), or intramuscular (i.m.) routes. Injection techniques are illustrated in Figures 3C and 3D. Intramuscular injections are done into the thigh muscle, with the mouse held in a similar way. Intravenous injections are done into the tail vein, with the mouse in a restraining device (Figure 3E). The i.v. route is probably best for stimulation of the spleen, s.c. and i.m. routes stimulating local lymph nodes preferentially. Cells injected i.p. very rapidly reach the bloodstream, and this route is probably almost as effective as i.v. and technically simpler. Emulsions in FCA should not be injected i.v. unless they have been solubilized, and this reduces the efficacy of FCA. The adjuvanticity of FCA depends, in part, on the formation of a water-in-oil emulsion which forms a depot at the site of injection, releasing antigen continuously. The emulsification of the antigen and adjuvant must be done properly to achieve maximal immunogenicity. Procedures for emulsification and testing of the emulsion are described in Mishell and Shigii[4] and by Herbert.[5]

*4. Booster Injections*

A single booster injection, given after the primary antibody response has subsided, will act on the cells which carry immunological memory and will lead to a secondary antibody response. A 4-week interval between primary immunization and booster is generally effective. The antibody response, especially against antigens which are only weakly immunogenic, may be increased by giving numerous booster injections. An example is provided by Stahli et al.,[6] who gave five booster injections in the week before hybridization, in order to stimulate a maximal number of blast cells in the spleen.

On the other hand, a booster injection may be omitted altogether, as shown by Trucco et al.[7] Direct intrasplenic immunization, discussed below, also relies on a single immunizing injection. The known differences between a primary and secondary response indicate that the hybrids obtained after a primary immunization are likely to be fewer in number and to produce antibody of lower affinity, which is also likely to belong to the IgM class. Certain types of antigens, which elicit a response largely independent of T lymphocytes, do not induce a secondary response. These antigens are typically long chain polymers with identical repeating structures, particularly polysaccharide antigens. In such cases there is unlikely to be anything gained from a secondary immunization.

The booster injection can consist of a lower dose of antigen than the primary, since the secondary response is more vigorous. Adjuvant is less helpful in the secondary response, and FCA may cause anaphylaxis. Freund's incomplete adjuvant is often used in booster injections, since it has some of the adjuvant action with a lower risk of anaphylaxis. The use of Bordetella pertussis as an adjuvant for the secondary response may be advantageous, since this adjuvant tends to elicit a large number of blast cells. Polyclonal B-cell activators such as lipopolysaccharide or pokeweed mitogen have also been used to elicit a blast cell response, but the majority of the blasts so induced will probably not be relevant to the antigen. Since the i.v. route is the one which stimulates the spleen most directly, this is the preferred route for booster injections.

*5. Improving Specificity*

Before the development of the hybridoma technique a number of approaches were used to direct the immune response to the production of antibody against particular components of a mixture. It is important to do this because the components of a mix-

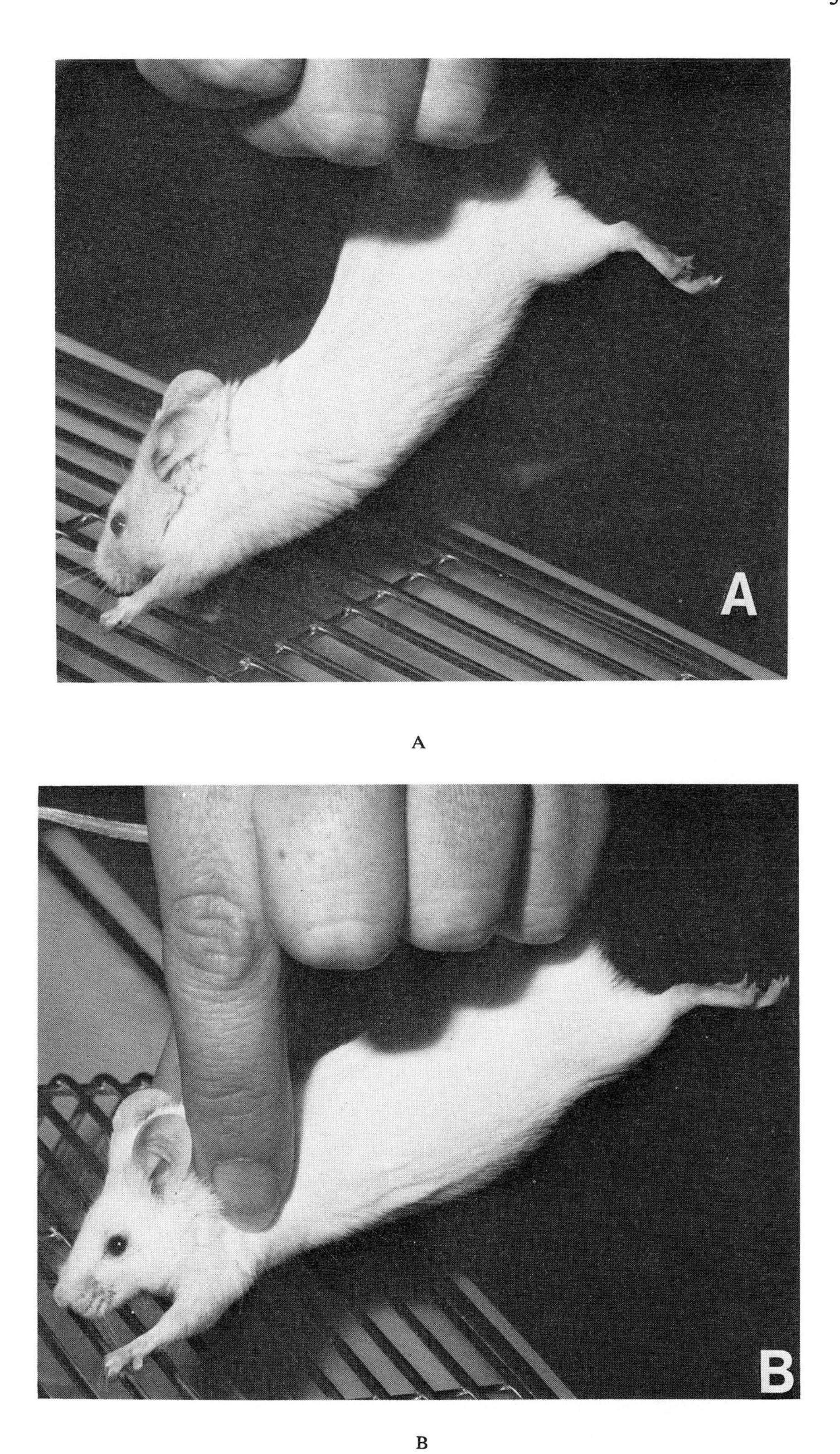


A

B

FIGURE 3. Injection methods. (A and B) Picking up and holding the mouse; (C) intraperitoneal (i.p.); (D) subcutaneous (s.c.); (E) intravenous (i.v.) injection.

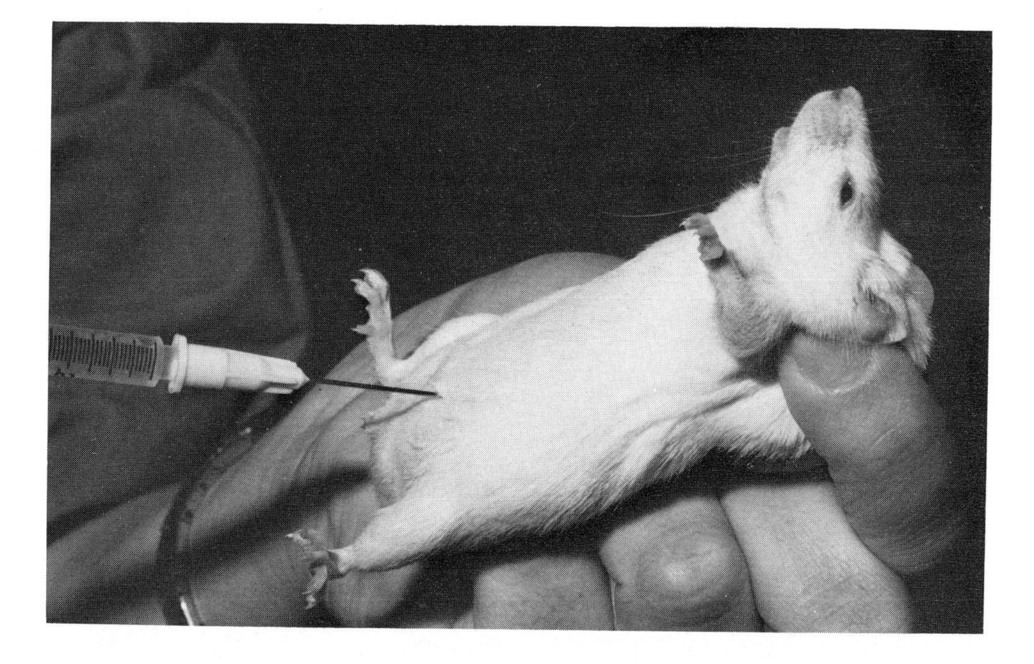

FIGURE 3C

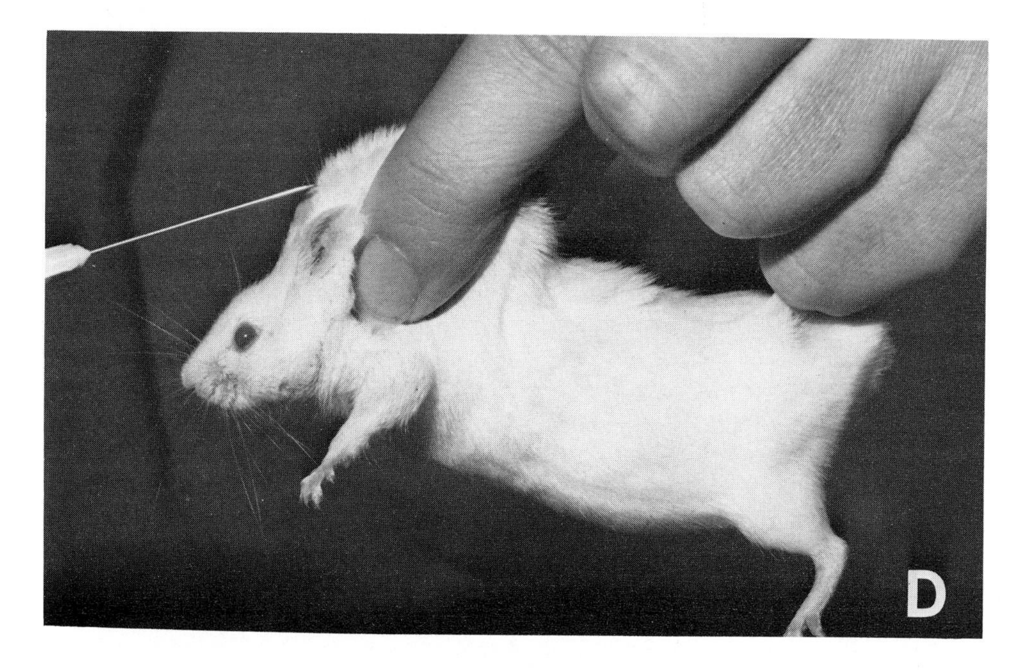


FIGURE 3D

ture differ in immunogenicity. In hybridoma production this means that the majority of hybridomas produced against an antigen mixture will be directed against "dominant antigens". Procedures used to point the immune response in the desired direction include induction of tolerance to the other antigens and reducing their immunogenicity

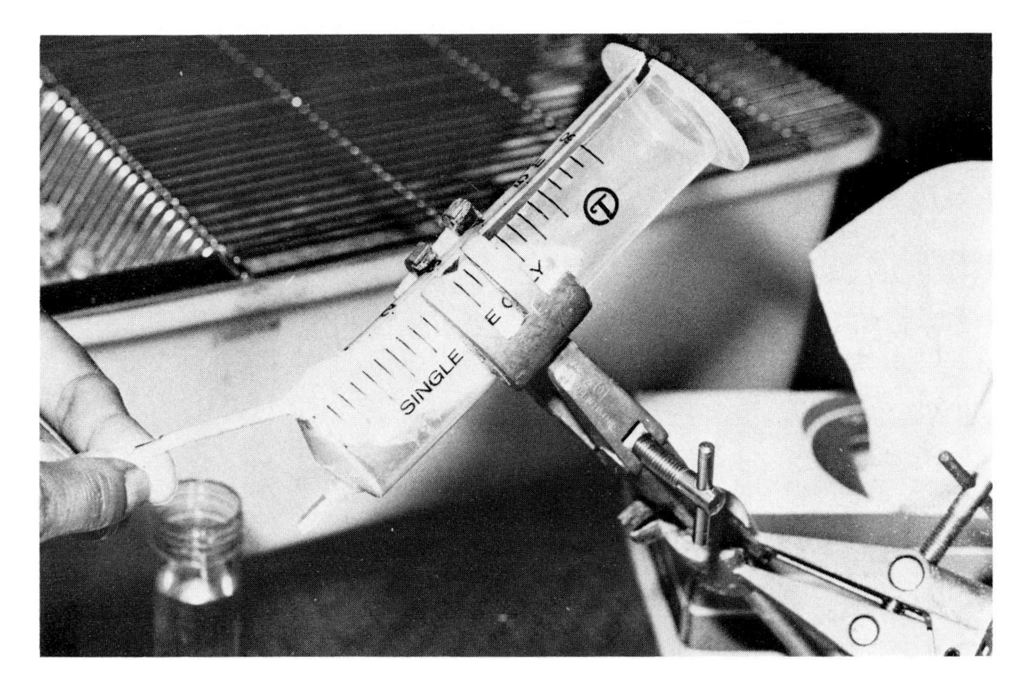


FIGURE 3E

by "coating" with antibody. The latter procedure is much easier and has been shown to be effective in increasing relative titers of polyclonal antisera against the desired antigen.[8] In hybridomas it may be expected that this procedure will lead to a greater proportion of hybrids being directed against the antigen of interest, and this procedure is used extensively.

### *6. Serum Antibody to Assess Immunization*

Measurement of antibody in the serum of the immunized mouse serves as a guide to successful immunization, particularly since individual mice can differ greatly in their response to the same antigen. When the antigen is in a complex mixture it may be difficult to determine the level of antibody against the antigen of interest, but an overall antibody titer against the antigen mixture at least serves to indicate that the mouse has responded.

### *7. Local Immune Stimulation*

A technique which has been recommended when the amount of antigen available is severely limited is direct injection of the antigen into the spleen.[9] This method also has the advantage of a very short immunization time, but is technically more complex. Another approach to localize stimulation may be the injection of small doses of antigen into the footpad. This generally leads to a marked enlargement of the popliteal lymph node, which is likely to serve as a rich source of antibody-producing blast cells.

## VI. HYBRIDIZATION

Hybridization consists of the preparation of two cell types (the myeloma line, containing genes which confer the ability to multiply indefinitely and secrete immunoglobulin, and the spleen cell, containing the genes coding for antibody against the immunogen) and the actual fusion process (Figure 1).

Table 1
MYELOMA AND LYMPHOBLASTOID LINES USED FOR B-HYBRIDOMA PRODUCTION

| Line | Features | Ref. |
|---|---|---|
| Mouse | | |
| P3-X63-Ag8 | Secretes $IgG_1$, now rarely used | 10 |
| P3-NS1/1-Ag4-1 | Hybrids secrete K light chain; still used extensively but should be replaced; HAT selection | 11 |
| P3-X63-Ag8.653 | No Ig synthesis; HAT selection; the most widely used nonsecreting line | 12 |
| SP2/0-Ag14 | No Ig synthesis; HAT selection | 13 |
| FO | No Ig synthesis; HAT selection | 14 |
| S194/5.XXO.BU.1 | No Ig synthesis; HAT selection | 15 |
| Rat | | |
| 210-RCY3-Ag1 | Hybrids secrete K light chain; HAT selection, Lou/C rat strain | 16 |
| YB2/3 Ag20 | No Ig synthesis, HAT selection, Lou/C rat strain | 17 |
| IR 983F | No Ig synthesis, HAT selection; Lou/C rat strain (different substrain) | 18 |
| Human | Several lines are available,[2,3] but none is accepted as routinely successful | |

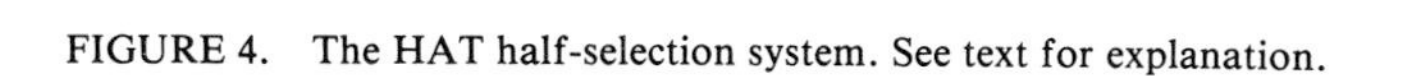


FIGURE 4. The HAT half-selection system. See text for explanation.

## A. The Myeloma Cell Line

The properties required of a myeloma line to be used for hybridization are as follows: the myeloma, and hybrids derived from it, must grow continuously in vitro; the myeloma must bear a genetic marker that enables selective growth of hybrids rather than unhybridized myeloma cells; the myeloma must confer on hybrids the ability to secrete immunoglobulin in quantity, but preferably should not code for its own immunoglobulin.

A number of suitable myeloma lines of mouse, rat, and human origin are available and are listed in Table 1. Much of the earlier work on mouse hybridization was carried out with lines which coded for their own immunoglobulin or immunoglobulin light chains. Hybrids produced from these myeloma cells secrete immunoglobulin molecules which consist of mixtures of the light and heavy chains coded for by the genes of the two parental cells. Much of this immunoglobulin does not react with the immunogen. Thus, myeloma lines which do not code for their own immunoglobulin are preferable.

The selection system used almost universally is the HAT system of Littlefield.[19] In this system (Figure 4) aminopterin (A in HAT) is used to block the main biosynthetic pathway for nucleic acid. Normal cells can continue to synthesize nucleic acid using the salvage pathway, provided they are supplied with hypoxanthine and thymidine (H and T in HAT). Mutant cells which lack one of the enzymes required for the salvage pathway cannot multiply in the presence of aminopterin, and die out. Mutant myelomas, lacking the enzyme hypoxanthine guanine phosphoribosyl transferase (HGPRT), have been prepared. These cells are used as the myeloma "parent" in the fusion and will multiply in the absence of aminopterin, but die out in HAT medium. Fusion products, which have the HGPRT gene from the normal spleen cell fusion "parent", survive in HAT medium. This system is referred to as half-selective, since only one of the fusion partners has a genetic defect and is selected against by the medium. Spleen cells will not grow in culture without special stimuli and conditions, so that there is no need to select against unfused spleen cells.

The HAT system works well and is, thus, used for the bulk of the work in this area. Alternative half-selective systems and doubly-selective systems may be needed for particular purposes. A cell line deficient in thymidine kinase would work equally well in the HAT system. Mutants lacking adenine phosphoribosyl transferase or adenosine kinase can be used in selective media. Selective resistance to amphotericin B methyl ester or ouabain can also serve as the basis of selection systems. An example of a doubly-selective medium which has been used for the production of hybrids between myeloma and lymphoblastoid cell lines, both of which can proliferate, is to make the same cell (the myeloma) HAT sensitive and ouabain resistant and grow the hybrids in HAT medium with ouabain. The myeloma cells die out because of the aminopterin; the lymphoblastoid cells die out because of their sensitivity to ouabain. Only hybrids carrying ouabain resistance (conferred by the myeloma parent) and HGPRT (from the lymphoblastoid line parent) can survive. These alternative systems may find specialized usage, but the bulk of mouse hybridoma work utilizes the well-established HAT system.

The myeloma lines are maintained in culture and should be growing actively at the time of fusion. Ampules of cells are stored in liquid nitrogen. The original lines have been distributed around the world and any individual culture would have a long and tortuous history. It is, thus, not uncommon to obtain a culture with the right name which does not produce hybrids. It is sensible to obtain the original stock from a laboratory which is routinely using the line successfully.

The myeloma lines have occasionally been reported to "revert" to cells which grow in HAT. If this happens, or as an occasional precaution, the cells may be selected by culturing in the presence of 6-thioguanine (40 $\mu$g/m$\ell$) or 8-azaguainine (20 $\mu$g/m$\ell$). These compounds are incorporated into nucleic acid by HGPRT and kill the cell. Thus, culture in either of these compounds will purge the population of any cells which have reverted to expressing the enzyme.

B. Spleen Cells

The spleen cells are derived from an animal immunized as discussed in Section V. A single-cell suspension is prepared using procedures which cause minimal trauma. The cells may then be used without any enrichment, after removing erythrocytes, or after selection of antigen-reactive cells. Any degree of enrichment is clearly beneficial, since it will increase the probability of the appropriate cells (B lymphocytes, probably B-cell blasts, in particular) coming into contact with the myeloma cells. For this reason the removal of erythrocytes is certainly recommended. Selective purification of antigen-reactive cells should be beneficial, and studies in model systems have confirmed this.[20] However, too little is known about the best methods for selection, the benefits, and the losses of cells to recommend specific enrichment as a routine method. It must also be remembered that specific enrichment as a routine method. It must also be remembered that specific enrichment will not necessarily increase the numbers of useful hybrids, only their relative proportion. If the screening assay can readily distinguish relevant from irrelevant hybrids, the benefits of selective enrichment will be minor. Nevertheless, selective enrichment should be considered as an option, especially as the methods are further established. Two novel procedures favor the fusion of the antigen-specific cells, rather than enriching these cells beforehand.[21,22]

The viability of the spleen cell preparation is a major determinant in the success of fusion, and care must be taken in preparing these cells. Methods should be simple, gentle, and rapid. Suitable methods are described in Section C.3 below.

## C. The Fusion Process

### *1. Introduction and Rationale*

When cells are brought into close contact, fusion of membranes will occur, but this will be a very rare event. The frequency of fusion can be increased by a variety of agents. The mechanisms are not fully understood. Sendai virus was extensively used in early hybridization studies, but polyethylene glycol is now used almost universally, since it is easier to use and more effective. The fusion process is inevitably traumatic to cell membranes, and the procedure should be designed to minimize this damage.

The variables in a fusion experiment are as follows: myeloma cell (suitability, viability), spleen cell population, cell ratio, medium (pH buffering, presence of serum), conditions for achieving cell contact, fusing agent and additives, conditions of fusion (time, pH, temperature, physical handling), and processing after fusion. Some of these variables cannot be controlled precisely, so that the operator becomes an additional variable. Two people hybridizing aliquots of the same cell mixture side by side can get different results, and experience seems to be beneficial. One useful indicator of how well the procedure has been handled is the viability of the cell mixture after fusion.

### *2. Materials and Preparation*

- Immunized mice (see Section V): it is generally adequate to fuse one mouse spleen at a time, and any attempt to scale up should be carefully evaluated, since the procedures described below have been arrived at empirically for single-spleen fusions.
- Myeloma cells in actively growing culture, split to 2 to 3 × $10^5$/mℓ the previous day.
- Complete medium.
- 24- or 48-well cluster plates (well size 2 or 1 mℓ) are placed in the incubator a few hours before use. This allows the plastic to reach 37°C and the air in the plates to be replaced by 5% $CO_2$ in air.
- 70% alcohol.
- Sterile surgical equipment: two pairs of small scissors and two pairs of small forceps, one pair preferably curved; after a few dissections readers will find their own favorite combination of instruments.
- 5- to 10-mℓ disposable sterile syringes and 23-gauge needles.
- Sterile tissue-culture grade petri dishes, 28 mm diameter.
- Sterile plugged Pasteur pipettes and 1- and 10-mℓ graduated pipettes, pipette aid (mouth pipetting is banned in most biological laboratories).
- 20- to 30-mℓ sterile centrifuge tubes with "V" bottom.
- Geys hemolytic medium.
- PEG/DMSO fusion solution.
- HAT medium.

Before starting ensure that the myeloma cells are healthy (viability > 90%, no contamination) and present in sufficient numbers (about 10 million/spleen). Wash the sterile working cabinet down with alcohol and leave the sterile air flowing for at least 30 min before starting. Arrange flasks of medium, surgical equipment, pipette cans, etc. in the cabinet, avoiding "shadows" (objects placed in such a position that any spores on their outside surface can be blown onto the sterile work). As an additional precaution, outside surfaces of objects placed in the cabinet can be swabbed down with 70% alcohol. This is particularly important for bottles which have been placed in a waterbath to warm up, because a 37°C waterbath is liable to be contaminated, even if disinfectant is added to the water.

For each spleen, place a tube containing just over 1 mℓ PEG/DMSO, one containing 3 mℓ complete medium, and another containing 10 mℓ complete medium in a 37°C waterbath adjacent to the working cabinet. Check that everything is available before commencing.

*3. Procedure*

1. Kill the mouse by cervical dislocation or $CO_2$ asphyxiation (by placing it in a closed container with a few pieces of dry ice). Swab the mouse liberally with 70% alcohol, or place the dead mouse briefly in a beaker of alcohol, before placing it in the working cabinet; fur blowing about in the cabinet is not conducive to sterility.
2. Make a small incision in the skin and tear the skin to expose the abdominal wall. Pull the skin well out of the way to avoid loose hair getting into the work. Using sterile scissors and forceps, make a small incision in the abdominal wall and take out the spleen (Figure 5). Place it in a petri dish containing 10 mℓ complete medium at room temperature. Remove fatty tissue adhering to the spleen.
3. Inject medium into the spleen, causing it to swell and cells to be released (Figure 5). Using sterile forceps and syringe needle, tease tissue apart. This should be done rapidly and gently, until further teasing apart of small fragments will not release many more cells.
4. Tilt the petri dish and use a Pasteur pipette to transfer the cells into a centrifuge tube, leaving clumps behind. If any clumps are transferred to the tube, allow them to settle and pipette the cells off into a fresh tube, leaving the clumps behind. Centrifuge for 5 min at 200 g. This process and subsequent steps in the preparation of the cells are carried out at room temperature.
5. While the cells are being centrifuged, make up the working strength Geys hemolytic medium: for each spleen 14.5 mℓ distilled water, 4.0 mℓ Geys solution A, and 1.0 mℓ Geys solution B. Adjust pH to 7.2 (by color of indicator). Sterilize by membrane filtration if any of the ingredients are not sterile. This medium should be used within 30 min of making up, because it is unstable. The medium should be at room temperature.
6. Remove supernatant liquid from the cell pellet, leaving behind as little medium as possible. Flick tube to loosen the pellet, and add the Geys hemolytic medium (the amount made up, i.e., just under 20 mℓ, to one spleen). Add a small volume first and resuspend the cells in it, either with a Pasteur pipette or by gently flicking the tube with a finger, then add the rest of the medium. Leave to stand at room temperature for exactly 5 min, then centrifuge for 5 min at 200 g. Resuspend cells in 10 mℓ complete medium at room temperature.
7. While the spleen cells are being treated in Geys hemolytic medium and centrifuged, prepare the myeloma cells. The cell count and viability will already have

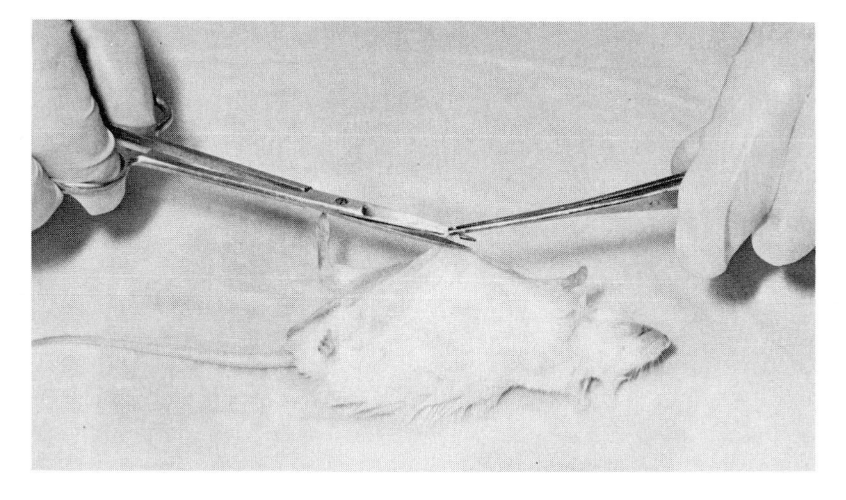

A

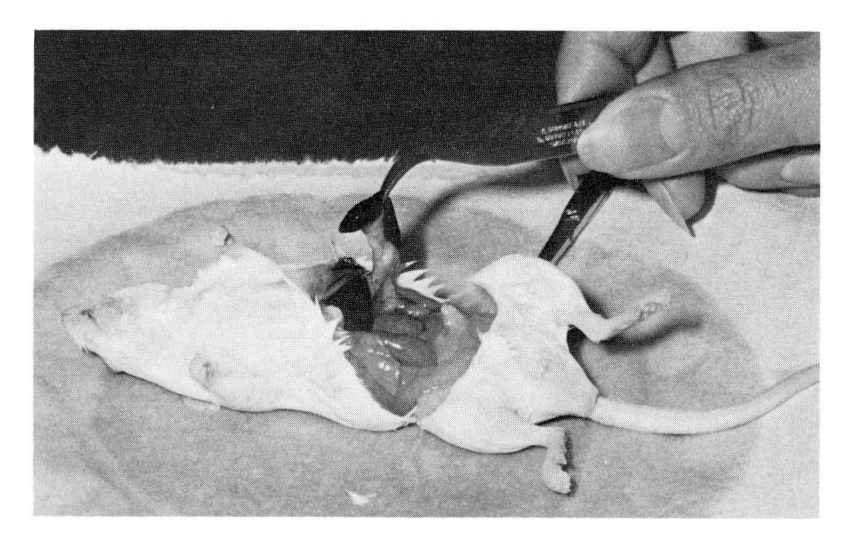

B

FIGURE 5. Removal of spleen and preparation of a single cell suspension. See text.

been determined. Take a volume containing $2 \times 10^7$ cells and centrifuge at 200 g for 5 min. This is conveniently done at the same time as the hemolyzed spleen cells are being centrifuged.

8. Carry out a cell count and viability determination on the spleen cells. There should be few red cells, and the viability should be >80%. Add a volume of myeloma cell suspension containing 10% of the total viable white cell content of the spleen cell suspension, i.e., the spleen cell/myeloma cell ratio is 10:1.
9. At this stage the cells are ready for hybridization. They may be left at room temperature for 30 min if needed for coffee break etc. Make sure the PEG/DMSO and medium aliquots are ready in the 37°C bath, and a sterile 1-mℓ pipette and stopwatch are available. When everything is ready, centrifuge the cell mixture in a 20- to 30 mℓ "V"-bottom centrifuge tube, for 5 min at 200 g. Remove the medium carefully and completely. This is best done by pouring off,

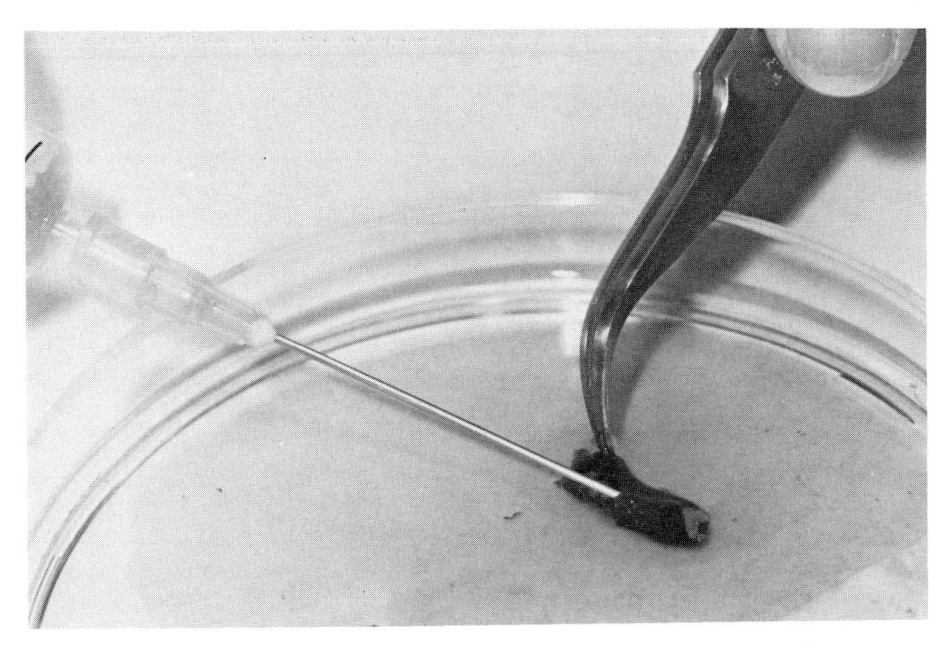

FIGURE 5C

FIGURE 5D

making sure to remove the drop that collects in the top of the tube. Flick the tube to loosen the pellet; pipette 1 mℓ of PEG/DMSO onto the cells, starting the stopwatch at the same time. Rapidly mix the cells in the viscous solution by stirring and pipetting up and down. The mixture will remain lumpy, but the clumps should be loosened by this treatment. After 1 min from the addition of the PEG/DMSO, start the dilution process.

10. Add warm medium from the 3-mℓ aliquot dropwise from a Pasteur pipette, mixing by gentle swirling. The 3 mℓ should be added over a 10-min period. Add the

10-mℓ warm medium over a further 10-min period, dropwise and with gentle mixing. If mixing is too vigorous the incipient hybrids, which are at this stage cells sticking to each other, may be separated.

11. Centrifuge at 200 g for 5 min. Resuspend gently in a small volume of complete medium, and dilute to 20 mℓ. Take an aliquot and determine the viability. Place the cell suspension in a sterile petri dish and put it in a 37°C gassed incubator. The cells can remain in the incubator for 1 to 3 hr, time for lunch and preparations for plating out. The incubation allows the DMSO to diffuse out of the cells.
12. Prepare the feeder cell suspension. Prepare spleen cells from an unimmunized mouse (of the same strain as the immunized mouse) in the same way as described for the immune spleen cells, including Geys lysis. Make up at $2 \times 10^5$/mℓ in HAT medium. The feeder cells may conveniently be prepared at the same time as the immune spleen cells, before fusion.
13. Resuspend the hybridized cells and transfer to a centrifuge tube, washing loosely adhering cells from the petri dish. Determine cell count and viability. Centrifuge cells at 200 g for 5 min and resuspend in HAT medium. Adjust the cell concentration, based on the *original* number of cells taken, to $4 \times 10^5$ *myeloma cells/mℓ*.
14. Mix feeder cell suspension and hybridized cell suspension in equal volumes. Plate out at 1 mℓ/well of the 24-well plates, or 0.5 mℓ/well in 48-well clusters. Place plates in the gassed incubator.
15. Relax!

*4. Notes*

The method described is based on the procedures published by Galfre et al.[23] There are many alternative procedures, some significantly different and others very similar. A method which is significantly different and widely used is that of Gefter et al.[24] The following notes attempt to pinpoint the important features of the process and the steps where variation of the procedure may be beneficial. It has to be emphasized that the mechanism of hybridization is not really understood, so that the procedures recommended have been arrived at empirically. It is very likely that these procedures can be improved on, but this should be done by careful experimentation, changing one variable at a time.

1. Spleen cells: the procedure must be gentle. Dye exclusion detects only severe damage to cells. Procedures such as forcing cells through a wire mesh give better yields but poorer viability. The use of Gey's hemolytic medium rather than simpler hemolytic solutions is recommended to reduce damage to the lymphoid cells. If the viability of the spleen cells before fusion is <80%, examine the viability after each step in the next hybridization experiment, to see at which stage cells are being damaged.
2. Cell ratio: the ratio of 10:1 (spleen/myeloma) is used by most laboratories. If the spleen lymphocytes are further enriched, it would be appropriate to reduce the ratio.
3. Fusion conditions: serum is not necessary in the medium and has been said to be detrimental. However, the viability of the cells is adversely affected in the absence of serum. RPMI medium is not an effective buffer outside a gassed incubator, and some workers add HEPES. HEPES is toxic to cells, although its extensive use in cell preparations shows that it can be used successfully. We prefer to work with RPMI medium, making sure it is not alkaline, as indicated by the color of the indicator (reddish/purple medium is alkaline, yellow medium is acid, and

orange medium is neutral [pH 7.2 to 7.6]) before starting, and working rapidly to avoid excessive loss of $CO_2$. HEPES and DMSO should not be used together because of increased toxicity. The comments of Westerwoudt[1] on the effect of pH on the successful outcome of heterokaryon formation are interesting. High pH (8.0 to 8.2) favors successful fusion, whereas at lower pH premature chromosome condensation, which leads to cell death, is observed. However, premature chromosome condensation appears to result only when cells in mitosis fuse with interphase cells.[1] Most myeloma and spleen cells will be in $G_0$ or $G_1$ phases of the cell cycle. We prefer neutral pH (7.2 to 8.0) in order to maintain cell viability.

4. Cell suspensions are best maintained at room temperature during preparation. The fusion process and subsequent dilution should be at 37°C, and we have on occasion carried out the fusion and dilution with the tube containing the cells standing in a small container of water at 37°C. However, this significantly increases the difficulties in keeping everything sterile, and it seems to be adequate to use solutions which are at 37°C at the outset.
5. DMSO is not essential in the fusion mixture, but seems to increase the yield of hybrids. Note that the presence of DMSO affects the indicator color, making it look more alkaline.
6. The actual fusion procedure is probably not ideal; certainly it is difficult to do exactly the same thing each time. Various other procedures have been recommended.[1,24] The conditions are intended to encourage fusion without causing excessive cell damage. The dilution process is critical; if it is too vigorous the yield of hybrids will be reduced. Cells are particularly fragile after the fusion, and it is important to keep centrifugation and resuspension steps to a minimum. On the other hand, it is important to reduce the concentration of PEG and DMSO. The recommended procedure is a compromise, with only a single wash immediately after dilution, followed by a longer period in medium, during which DMSO can diffuse out of cells and the fusion process can continue in cells which have started to fuse.
7. Feeder cells: these are not essential, and the fusion mixture contains nonimmune spleen cells already. However, the feeder cells have not been through the damaging fusion process, and in practice results are significantly better if feeder cells are used. Thymocytes,[25] macrophages,[14] and human fibroblasts[26] can be used as feeder cells. If the feeder cells are capable of dividing (fibroblasts, for example) they should be irradiated to prevent them overgrowing the hybrids. An alternative to feeder cells, which also promotes the growth of colonies, is supernatant from human endothelial cell cultures.[27]
8. Plating out: most laboratories use 96-well microtiter plates, each well having a capacity of 200 $\mu\ell$. This yields a higher number of cultures per spleen, but if many of them have no colony growth this is not a real advantage. We have repeatedly tried to use microplates, but have concluded that our success rate is higher with the larger (1 or 2 m$\ell$ capacity) wells. One important reason is that the larger wells can accommodate larger colonies, which makes for greater flexibility when screening colonies. If, using the recommended conditions, mutiple colonies are regularly found in most wells, the concentration at which the hybridized cells are plated out should be reduced. If one or fewer colonies per well are obtained, it is unlikely that microwells would produce a greater yield of colonies. The 24 well plates are shown in use (Figure 6) and being examined on an inverted microscope (Figure 7).

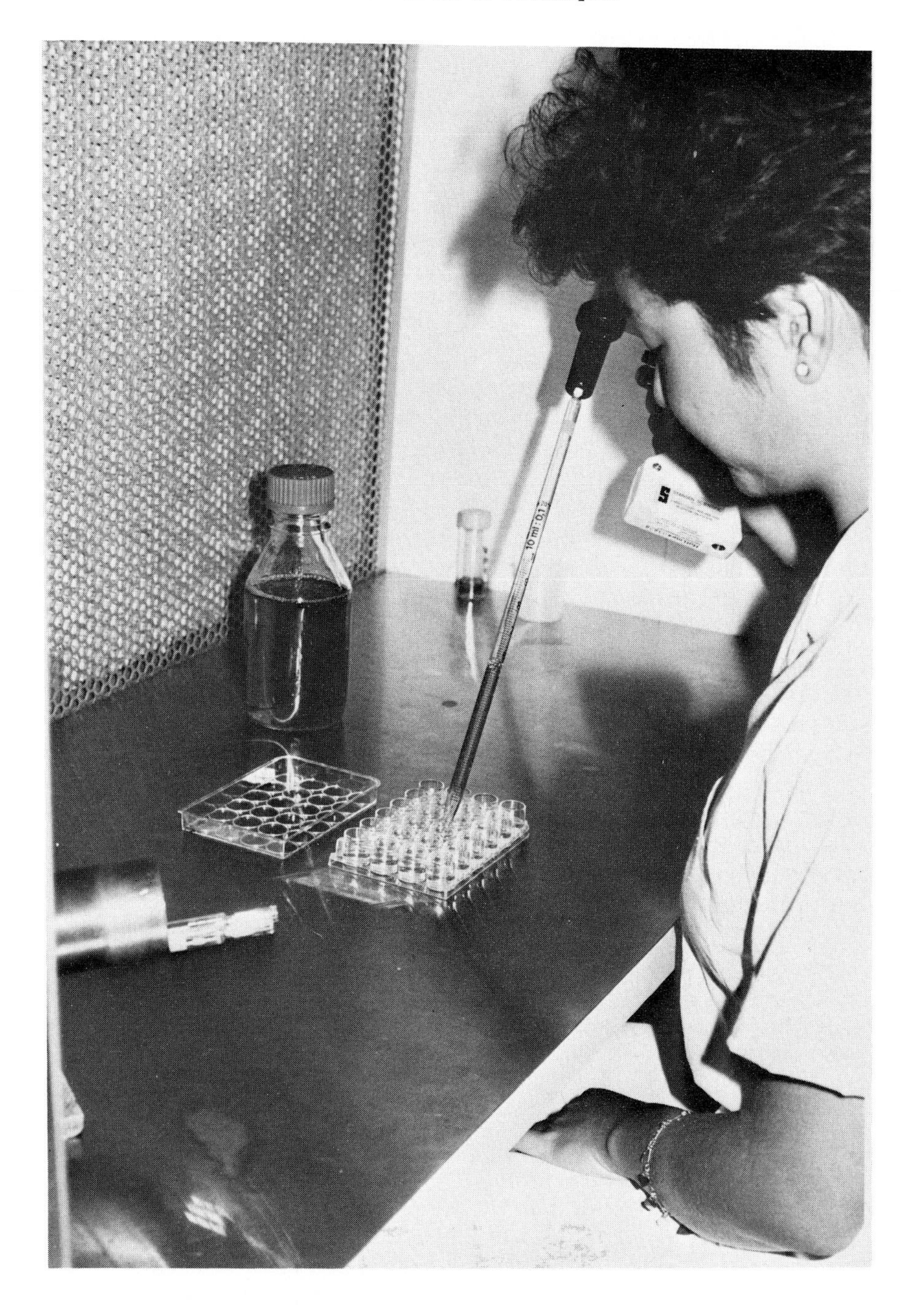

FIGURE 6. Cell culture: cells being dispensed into a 24-well cluster dish. The work is carried out in a laminar flow cabinet, using sterile plugged pipettes and an electric pipette-aid, which avoids the need for mouth pipetting.

## VII. MAINTENANCE AND PROPAGATION

### A. Introduction and Rationale

This section will cover the tissue culture procedures, starting after the plating out of the fusion mixture, up to the point when screened colonies are ready for cloning (Fig-

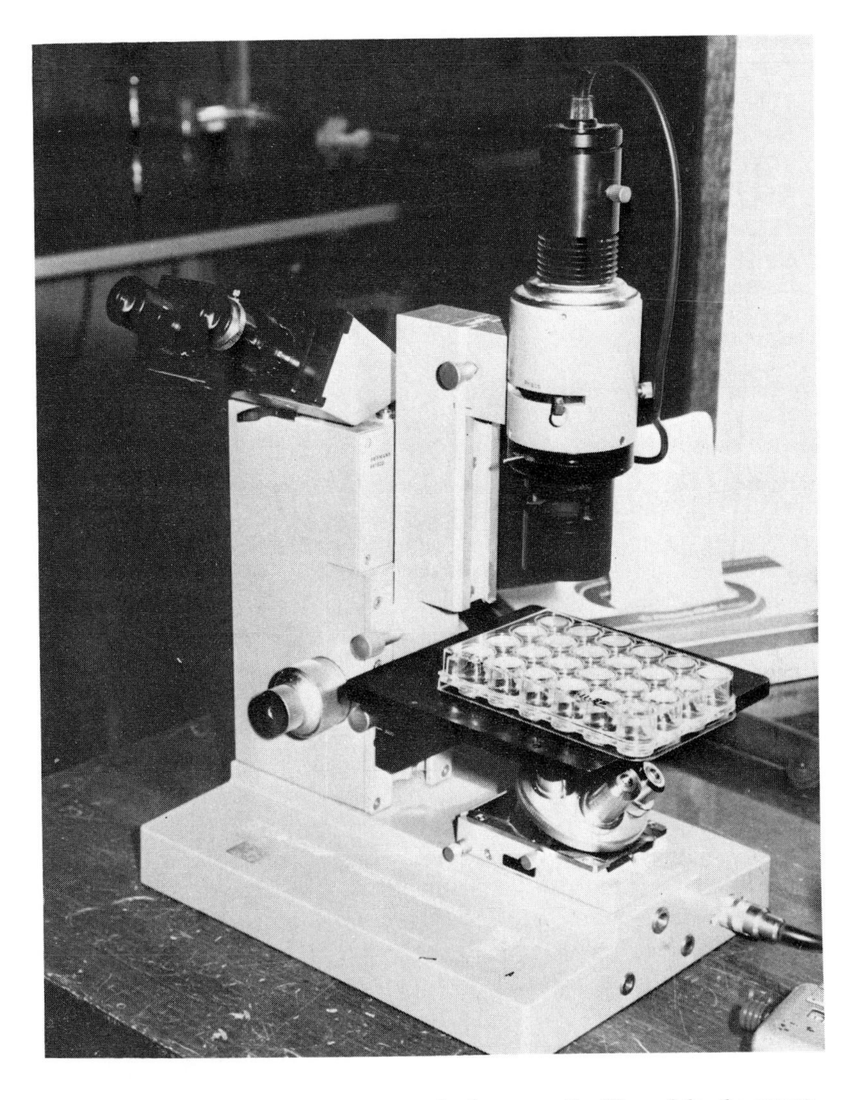

FIGURE 7. A 24-well dish on the inverted microscope. See Figure 8 for the appearance of cells in the inverted microscope.

ure 1). The objective during this phase of the work is to encourage growth of hybrids, to prevent growth of anything else, and to maintain hybridoma colonies while their secreted antibodies are tested. This phase of the work is no less important than the fusion process itself. Preliminary studies have indicated that the majority of fusion products detectable immediately after fusion die out in the next few days.[28] Thus, it is likely that good technique during the propagation phase will have at least as much influence on the yield of hybrids as good fusion technique. Hybrids do not start to divide immediately in culture, but have a variable quiescent period. Precisely what triggers hybrids to start dividing is not known, and many cells may die without reaching the growth phase. Unfused myeloma cells die out within the first week because of the aminopterin block, while the majority of the spleen cells die because they do not have the inherent capacity to divide or stay alive in culture without exogenous stimuli. Some macrophages and fibroblast-like cells adhere and divide slowly over the first week or two. Generally, these grow much less rapidly than the hybrids and do not cause any problems.

It is quite likely that in this complex environment the new hybrids are on a knife edge, and slight differences in technique or in ingredients may influence the yield of hybrids dramatically. Established hybrid cell lines will tolerate conditions which new hybrids probably will not.

A mouse diploid cell has 40 chromosomes; thus, a hybrid of two cells will have 80 chromosomes initially. Such a genome is unstable and chromosomes are lost over the first few days in culture. Loss of certain chromosomes will lead to cessation of growth. Loss of chromosomes involved in the synthesis or secretion of immunoglobulin will result in nonsecretor hybrids, a point which we will come back to later.

### B. Culture of Hybrids after Fusion

1. After hybridization, the plates are examined every 1 to 3 days to check for contamination, pH (color of medium), and colony growth. They should be returned to the incubator rapidly, and the incubator should not be opened longer or more often than necessary. It is good practice to reserve one incubator for cultures which do not need to be looked at very frequently, so that the $CO_2$ level does not fluctuate excessively.
2. Hybrids are fed 7 days after the initial fusion. Add 1 mℓ fresh HT medium to each 2-mℓ well (there is no need to supplement the aminopterin at this stage). The medium should be fresh, so that the glutamine level is high, and should be warmed to 37°C. Add the medium gently so as to avoid disturbing colonies, running it down the wall of the well. It is necessary to compromise; if the medium is added so slowly that by the time a plate has been fed the medium is turning alkaline the feeding is taking too long. Feeding a 24-well plate will take 5 to 10 min. If the colonies are fed with normal medium (rather than HT) at this stage, hybrids will die out. This is because the hypoxanthine and thymidine of the original medium are metabolized, while the aminopterin remains in culture. Thus, the main nucleotide biosynthetic pathway is blocked and the salvage pathway is starved of the necessary ingredients.
3. Hybrids are checked daily after the day-7 feed, looking for colonies and for yellowing of the medium. This is caused by the release of acid metabolites by cells, which may be hybrids, macrophages, or fibroblast-like cells, or revertant myeloma cells. Contaminating yeast or bacteria will also yellow the medium, but are readily identified. Mycoplasma infection is much more difficult to detect, but causes yellowing of the medium when the cell concentration is still low. Hybridoma colonies have a very characteristic appearance under the inverted phase contrast microscope (Figures 8A and 8B). The cells are very round and large. Occasional colonies of cells with a less regular appearance are seen, often spreading fibroblast-like cells (Figure 8C). These colonies have a more diffuse appearance.
4. When the medium is yellowing and the colonies are visible macroscopically, they are fed by removing 1 mℓ of medium (for 2-mℓ wells) and replacing it with 1 mℓ fresh HT medium. The feeding operation must be carried out gently to avoid disturbing the colony (see 2 above).
5. Screening: the object of the screening process is to identify interesting hybrids and reject negative or uninteresting wells, thus, concentrating the effort on positive hybrids. Start collecting supernatant samples when the medium is yellowing and the colonies are visible macroscopically. The sample is collected as part of the feeding operation. If the sample is taken too early a negative result will not be definitive. On the other hand, the colonies should not be allowed to overgrow

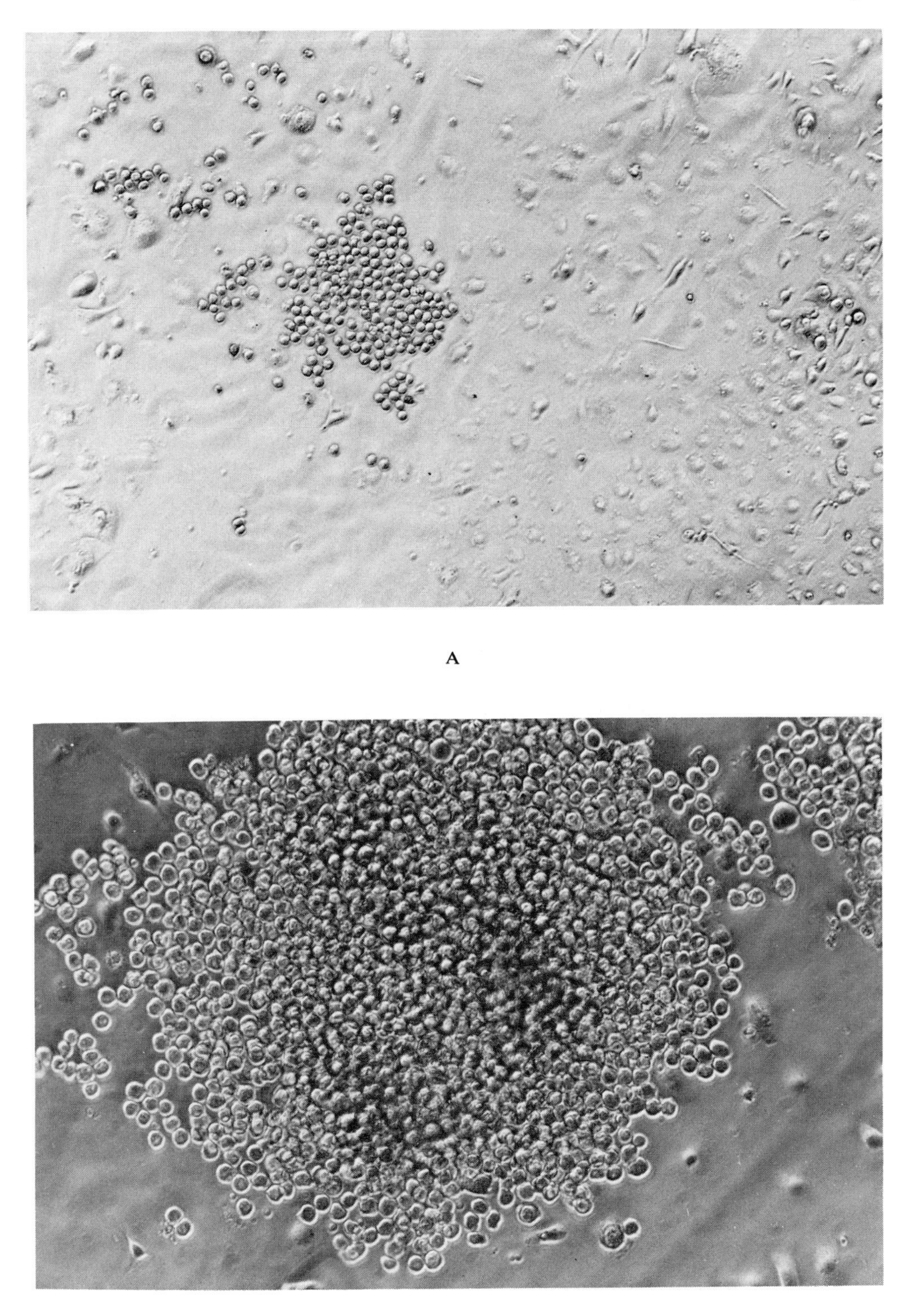

A

B

FIGURE 8. Appearance of cultures under the inserted microscope. (A) Small and (B) large hybridoma colonies; (C) colony of spreading adherent nonhybridoma cells.

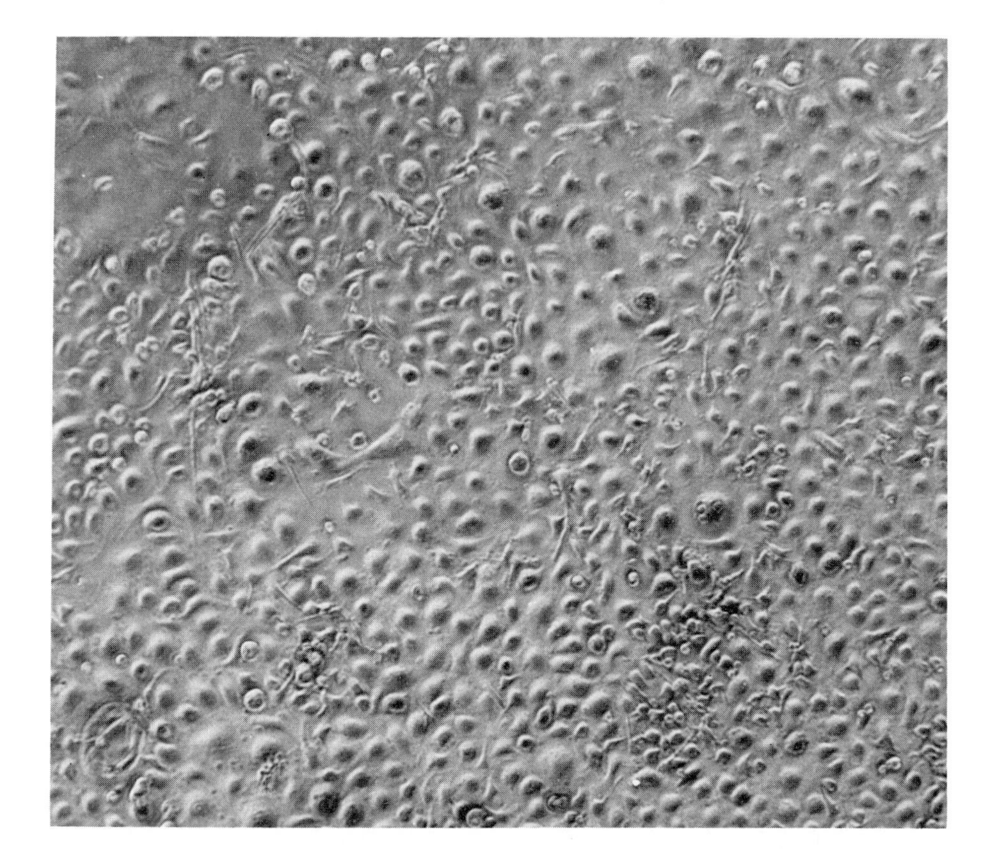

FIGURE 8C

or the medium allowed to become very acid, because cell viability will suffer. Our procedure, to allow for both these factors, is to collect samples when the colony is about 25% of the well area, and 2 to 3 days after the last feed. Wells which give negative results are retested when the colony is larger and the medium more acid, to make sure they really are negative. Wells which have been scored negative twice are excluded from further screening. Wells which are positive are cloned as soon as possible (see below). The screen may be a double screen to exclude positive wells with irrelevant reactivities (see Section VIII). Occasionally, a weak positive result may be obtained early and not repeated on retest. This may result from antibody secretion by the unhybridized lymphocytes. This is not a common problem, but if many wells have weak reactivity feed the wells and retest them when the medium turns yellow again; any activity found now is likely to be from hybrids. A number of automated sampling and feeding systems have been marketed in recent years, to reduce the labor component of hybridoma work. However, colonies grow at very dissimilar rates, and it is never necessary to sample a whole plate, or most wells of a plate, at one time. It is rarely necessary to feed the whole plate at one time, once colony growth is established in some wells. These considerations render the benefits of such automation doubtful.

6. Picking colonies. The wells which have been selected on the basis of the screen may have single or multiple colonies. Multiple colonies may result from disturbance of a single original colony. Even if the cells all come from a single colony they are not necessarily identical, because of the problem of chromosome loss

referred to previously. These considerations suggest that the best thing to do at this stage is to gently resuspend the entire well contents and remove them for cloning. If there appear to be two to three discrete colonies it may be worthwhile picking these individually, with a Pasteur pipette, and cloning separately. If the nonhybidoma cells (fibroblast-like or macrophages) are growing rapidly it may be advisable to pick the hybrid cells gently, avoiding transfer of the feeder cells. Because cloning is not invariably successful (see Section IX) two precautions are worth taking: (1) maintain the well by feeding it after removing most of the hybrid cells and (2) expand the uncloned population by placing in two to three wells (feeder cells should not be necessary provided the cell concentration is kept above $10^4/m\ell$). These cells may be further expanded and cryopreserved, and further cloning can then be carried out at any time in the future. Cryopreservation of these uncloned cells is not essential, since all subsequent work will be done on cloned cells. It provides a safety net, however, and the uncloned cultures should at least be maintained until the cloning plate is showing vigorous colony growth.

## VIII. SCREENING

### A. Introduction and Rationale

A screening test should identify colonies producing antibody against the immunogen. It should be rapid and capable of handling 20 to 30 samples at a time. The screen should be appropriate to the intended uses of the antibody, a point which will be amplified later. A double screening assay may be utilized to reject antibodies against uninteresting components of the immunogen. A preliminary screen for secretion of immunoglobulin may be utilized, but at least in mouse fusions this screens out very few colonies and is not worth the extra work.

In principle, any assay for antibody against the antigen of interest can be used for screening, and the worker familiar with the particular antigen will be in the best position to make a suitable choice. However, the special properties of monoclonal antibodies make certain types of assay unsuitable and other assays may need to be modified. In this section factors affecting the choice of screening assay will be discussed, and the principles of the major assays will be described. Some of the methodology is described in greater detail in Chapters 5 and 6.

#### *1. Primary and Secondary Interactions*

An assay which directly detects the binding of antibody to antigen is said to detect primary interaction. An assay which measures some further reaction of the antibody after it has bound to antigen is said to detect a secondary interaction. Examples of secondary interactions are complement fixation, agglutination, and precipitation. The distinction is important because secondary interactions depend on the class and subclass of antibody, and in some instances depend on the presence of several different antibodies. A comparison of two commonly used assays for antibody against cell membrane antigens, cytotoxicity, and immunofluorescence will illustrate the point. Cytotoxicity requires binding of the serum complement proteins. IgM and mouse IgG2a and 2b bind complement well, whereas IgG3 binds complement less effectively, and IgG1 does not fix complement at all. Thus, cytotoxicity is not a good general screen; it will miss many positive clones. On the other hand, if the antibodies are particularly required to be cytotoxic, cytotoxicity is the assay of choice (but must be carried out in a way which allows for the prozone phenomenon, see below). In immunofluorescence the antibody is reacted with the immunizing cell, and binding of the monoclonal antibody is detected using fluorescein-labeled antibody against mouse immunoglobulin (or other species as appropriate). This is, in principle, still a secondary interaction, but, provided the antiimmunoglobulin detects all subclasses of mouse immunoglobulin, indirect immunofluorescence can be considered as effectively detecting antibody binding, i.e., primary interaction. Other assays which may be considered as effectively detecting primary interaction utilize radiolabeled or enzyme-labeled antiimmunoglobulin.

*2. Single and Multiple Determinants*

Antibodies react with macromolecules or particulate antigens, but the actual binding site (epitope, antigenic determinant) is relatively small. A complex consisting of a soluble macromolecule with a single antibody molecule bound to it will usually remain soluble. Precipitation results when several antibody and antigen molecules bind together to produce a large aggregate. If the antigen has repeating determinants (for example, a sugar in a particular conformation in a polysaccharide) several antibody molecules will bind, and precipitation is likely. If the antibody is IgM, it will bind several antigen molecules and precipitation is likely. If the antigen lacks repeating determinants and the monoclonal antibody is not IgM, the antigen-antibody complexes will be small and unlikely to precipitate. In a polyclonal antiserum there will probably be antibodies against several different determinants on the antigen, a large complex will form, and precipitation will result. Thus, for antigens which lack multiple repeating determinants (and this includes most proteins) monoclonal antibodies do not precipitate well, and precipitation is not a good screening assay.

Cytotoxicity suffers from a similar problem. In this case the cell does have multiple copies of the same epitope, but the epitope density may be too low to allow effective complement binding and consequent cell death.

*3. The Prozone Phenomenon*

If a divalent or polyvalent antibody is mixed with cells bearing the antigen, antibody molecules will form bridges linking cells together and cause agglutination. However, at very high antibody concentration the most probable binding reaction will involve a separate antibody molecule reacting with each antigenic determinant, in other words, the antibody molecules will behave monovalently and there will be no bridges formed, no agglutination. This means that as the antibody concentration is reduced agglutination first increases, then decreases as the amount of antibody becomes too small to form bridges. This tiration curve is shown in Figure 9. Loss of reaction at high dilution is common to all immunological reactions, but loss of activity at high antibody concentration is not universal and is given a special name — the prozone phenomenon. The region of antibody excess is called the prozone. Prozones are found in agglutination, precipitation, and cytotoxicity, but are not generally found in primary interactions.

The practical significance of the prozone phenomenon in screening for monoclonal antibodies is that a negative result may simply mean that the antibody concentration is too high. Thus, assays have to be carried out at several dilutions. Since monoclonal antibodies can have very high titers, they can also have very extensive prozones. Assays which exhibit the prozone phenomenon are generally not suitable for screening assays.

*4. Sensitivity and Background Staining*

Immunological assays differ widely in sensitivity. As already stated, the screening assay should be appropriate to the intended use of the antibody. For example, a highly sensitive radioisotopic assay may detect a few colonies which would be missed by indirect immunofluorescence. However, if the intended use of the monoclonal antibodies

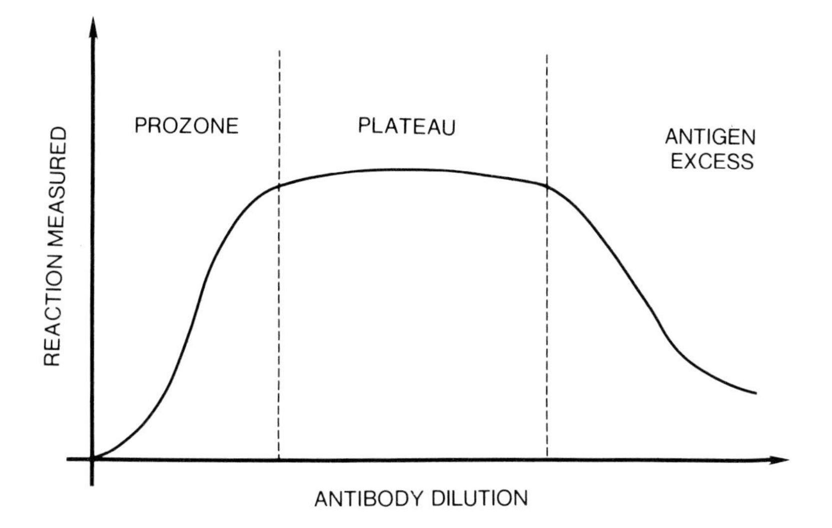


FIGURE 9. The prozone phenomenon. The test may be negative either because there is too much antibody (prozone) or because there is too little.

is in the identification of lymphocyte subsets in blood, the procedure which will be used in the routine laboratory is indirect immunofluorescence, and an antibody which does not perform well in this assay will not be useful.

Background, or nonspecific, binding always occurs to some degree in immunological assays and, strictly speaking, sensitivity refers to the ability of a test to discriminate between specific and nonspecific binding. Nonspecific staining has a number of sources and may involve either nonspecific binding of the monoclonal antibody or of the subsequent reagents of the assay. The most intractable background problem arises when making antibodies within a species, such as alloantibodies, autoantibodies, or tumor-specific antibodies. The problem here is that immunoglobulin is a normal constituent of tissues, particularly of B cells, and of cells and tissues in close contact with blood. The antiimmunoglobulin, which is an essential component of most screening tests, will react with this endogenous immunoglobulin, making it very difficult to detect binding of monoclonal antibody.

Other background problems arise when cells have Fc receptors enabling them to bind immunoglobulin, either the monoclonal antibody or the antiimmunoglobulin. Many cells have Fc receptors, particularly lymphoid and myeloid cells. Lesser background staining results from the natural "stickiness" of proteins. Screening assays require a "cutoff" value to discount background staining, and this can be set by using appropriate negative controls. Discrimination between weak positive staining and background staining is arbitrary, since most screening assays are not quantitative. A useful rule of thumb is not to try too hard to discriminate the weak positives; if they are not much stronger than the negative control they are unlikely to prove to be satisfactory reagents. This rule cannot be applied universally, however, because some important antigens are found at low concentrations, on cell membranes, for example. If there are only low numbers of molecules of the antigen the intensity of staining with antibody will inevitably be low.

*5. Speed and Convenience*

One final consideration in choosing a screening test is that it should be rapid and convenient. The test should be able to handle at least 20 to 30 samples, should be capable of completion and analysis within a day, and should be easy to interpret. The reason for this emphasis on speed and convenience is that answers are required quickly, in order to clone positive colonies as early as possible, before nonsecretors overgrow the secretors (this will be discussed in more detail in Section IX).

### B. Specific Applications and Examples

#### *1. Particulate Antigens*

Particulate antigens are often cells, and the assay chosen will depend on answers to the following questions. Is the antigen of interest expressed on the cell membrane or is it an internal component? Is the cell population uniform in its expression of the antigen, or are we looking for a monoclonal antibody which will only react against a subpopulation of cells?

Antibodies against membrane antigens on cells which are available in suspension may readily be detected using "second antibody" (antibody reacting against mouse immunoglobulin) which is derivatized either with a dye (for instance, fluorescein) with an enzyme (for example, alkaline phosphatase), or with an isotope (for example, $^{125}I$). The difference between using isotopic and colored label is that the isotopic methods are best suited to giving an estimate of total antigen content of a cell population, whereas methods such as fluorescence are suited to the analysis of cell mixtures. This is an important difference. A particular count from an isotope-binding assay can indicate either that 100% of the cells have X copies of the immunogenic determinant per cell, or that 20% of the cells have 5X copies per cell, the rest being negative. Immunofluorescence, on the other hand, tells us what proportion of cells in the mixture react with the monoclonal antibody.

When the cells are not available in suspension, the tissue must be sectioned for analysis. Immunofluorescence can then be used to detect the antigen, but color-forming immunoenzymatic procedures (immunoperoxidase or immunoalkaline phosphatase) are generally preferable because they allow counterstaining to delineate the structure of the cells and tissue. When the antigen of interest is an internal component of the cell, the antigen must be exposed either by cutting sections or by rendering the membrane permeable to antibody. Analysis is then conveniently done on tissue or, if the cells are in suspension, on cytocentrifuge preparations, by the immunoenzymatic methods.

##### *a. Indirect Immunofluorescence*

The procedure for indirect immunofluorescence is described in detail in Chapter 5 (Section II.A). This technique is particularly appropriate for screening hybridomas against surface antigens of cells in suspension, as discussed above.

##### *b. Staining of Tissue Sections, Cell Smears, or Cytocentrifuge Preparations*

Hybridomas against intracellular components, or against cells in tissue, can be screened using immunohistochemical methods, such as the immunoperoxidase method. The methodology is described in detail in Chapter 5, Section III. The preparation and screening of monoclonal antibodies for use on tissue sections is reviewed by Mason.[29] The methods are suited to the detection of antibodies to rare cell types, and are particularly useful when tissue localization of the antigen is an important criterion in determining whether the antibody is directed against the antigen of interest. This information would be lost if the screening were carried out by ELISA on a tissue homogenate. Tissue fixatives invariably destroy a proportion of antigenic determinants, while the use of cryostat sections of unfixed tissue entails the risk of loss of antigen by diffusion. These factors are considered in more detail in Chapter 5, Section

III.B. Additional information on specificity may be obtained by using composite slides with sections of several different tissues stained simultaneously.

*c. Cell ELISA, Immunofixation, Isotopically Labeled Antiimmunoglobulin Methods, and Rose-Bengal Assay*

This group of methods shares the common feature that cells are immobilized for the assay. In the Rose-Bengal assay[30] test antibody is immobilized on microwells which have been previously coated with protein A or antiimmunoglobulin. The target cells are then added to the plate; binding of cells is detected by staining them with the dye Rose-Bengal. The bound color is measured with an ELISA plate reader. The technique is rapid and automated; it requires little material. As with the other methods in this group, antibody against minor subpopulations may be missed. The Rose-Bengal method is clearly suited only to the detection of membrane antigens.

In cell ELISAs, cellular material is fixed to ELISA plates and the hybridoma supernatants are added. Binding of antibody is detected by the addition of enzyme-conjugated antiimmunoglobulin. There are a number of variants in the detection system, such as the use of biotin-labeled antiimmunoglobulin followed by avidin and biotin-labeled enzyme.[31] Immunofiltration is a variant on cell-ELISA in which the cells are trapped on a filter paper placed at the base of a microtiter plate which has had the well bottoms removed.[32] In either immunofiltration or the cell-ELISA, the enzyme-based detection system may be replaced by radiolabeled antiimunoglobulin. These methods are not intrinsically different from the corresponding techniques for soluble antigens, discussed in the following section, except that binding of cells or cell fragments to the ELISA plates is less reproducible than is the binding of soluble antigen. The larger the cell, the more difficult it becomes to obtain stable binding, and it may be necessary to fix the cells after binding. This is likely to lead to the loss of some antigenic determinants. Furthermore, it is important to ensure that the enzyme used to detect binding is not also present in the cells, since this would lead to false-positive results. In this regard, isotopic methods are preferable. If the cells are lysed before being fixed to the solid surface, these methods are not restricted to surface membrane antigen detection.

As has been pointed out, the assays in this group are not suited to the detection of antibodies which react with a small fraction of the target cell population. In the author's experience this has always been a major consideration and has led to the predominant use of the methods described in Sections a and b above.

*2. Soluble Antigens*

The most straightforward approach to the screening of antibodies against soluble antigens is to bind the antigen to a solid support, add test culture supernatant, and detect binding of the hybridoma species immunoglobulin. This is most often carried out as an ELISA, or using antiimmunoglobulin labeled with radioiodine. However, if the antigen has a measurable biochemical activity (for example, if the antigen is an enzyme), screening can be based on measurement of that biochemical activity. The hybridoma supernatant can be tested to see if it removes activity from solution. Alternatively, supernatant can be immobilized onto a solid matrix and mixed with the antigen, to determine if the biological activity is then "fixed" by the matrix. The latter reaction depends on the antibody not interfering with the biochemical activity. The relative merits of these three approaches have been discussed by Harris.[33]

It is important to remember that a positive test by ELISA or radioisotopic assay tells us only that we have antibody against a component of the mixture used in the assay. This component of the mixture is not necessarily the antigen of interest. Consequently, the purest available antigen preparation should be used in screening. This point may seem obvious, but is emphasized because the author has found it necessary to explain it on more than one occasion.

The methodology of ELISA assays is described in detail in Chapter 6, Section II.B.

*3. Commercial Screening Reagents*

The rapid growth in hybridoma work has led to the development of commercial screening reagents and kits which allow the nonimmunological laboratory to concentrate on the development of monoclonal antibodies and their applications, without the need to divert attention to the preparation of ancilliary reagents. This is a welcome improvement when contrasted to the situation a few years ago, when it was difficult to obtain even fluorescein-conjugated antimouse immunoglobulin of reproducible quality. Many immunological laboratories will still prefer to prepare their own reagents because of the cost savings; the decision will depend on factors such as the available facilities, volume of reagents needed, and costs. Commercial systems range from single reagents (antimouse immunoglobulin or a conjugated derivative being the single most important reagent needed), through reagent kits which incorporate detection reagents such as biotinylated antimouse immunoglobulin and avidin-enzyme conjugates, to complete screening systems. The screening systems are designed to harvest and test an entire 96-well plate. As indicated previously, it is seldom necessary to screen every well in one run, since the colonies grow at widely different rates. It may, nevertheless, be economic to screen the entire plate each time if the screen is readily automated, for example, an ELISA assay.

## IX. CLONING

### A. Introduction and Rationale

Once colonies have been identified as producing antibody of interest, the contents of the well (or the colony, if a colony is picked out of a well) should be cloned as early as possible. The objective of cloning is to ensure that the cells producing antibody comprise a monoclonal population. In the original fusion well the hybridoma cells may be descended from several fusion products. Furthermore, the gradual loss of chromosomes during the first days after fusion generates additional heterogeneity. In particular, nonsecreting loss mutants appear and may eventually outgrow the secretors. Although this happens most frequently soon after fusion, it can happen later, at any time in the culture of a hybridoma. Recloning may be necessary at any time, if the titer of antibody produced is found to be falling. Some workers recommend cloning each hybrid twice; we clone once and reclone if we have a reason to reclone — if the frequency of wells showing colony growth is higher than expected (see below), if the specificity of the antibody suggests it may not be monoclonal (see Chapter 4, Section IV.C.5), or when the antibody titer falls.

### B. Methods

In essence, cloning involves the setting up of single-cell cultures, each cell intended to grow into a colony of identical cells. Three different approaches may be used: limiting dilution in liquid culture, semisolid agar cultures, and selective isolation of antigen-reactive cells using a fluorescence-activated cell sorter.

*1. Cloning by Limiting Dilution*

***a. Rationale***

In this method, a suspension of hybridoma cells is plated out in individual culture wells at a dilution such that, statistically, the most probable number of cells in any particular well is 1. Poisson distribution statistics indicate that if the most probable cell number per well is 1, then 37% of wells will have no colonies at all. In practice, cells have a cloning efficiency which is less than 100% and variable. In other words, it may be necessary to plate out at more than one cell per well to obtain one colony per well. This leads to a degree of uncertainty as to the monoclonality of the resulting colonies. If, in a series of cultures started with the same inoculum, some have single colonies and some (37% or more according to Poisson statistics) have none, it is probable that the cultures with single colonies started from single cells. If the result of cloning is colony growth in the majority of cultures, it would be unwise to interpret a culture containing a colony as being derived from a single cell. If this happens, it is necessary to reclone, at a lower starting concentration. In order to maximize the chances of obtaining satisfactory single-cell clones with cells of unknown cloning efficiency, cultures are set up at several concentrations, 1, 3, and 10 cells or 0.5, 2, and 5 cells per well will generally give the desired result at one of these levels.

It is generally adequate to expand and cryopreserve only two to three positive clones, although more may be needed if there is reason to think that the original cell suspension contained more than one secreting cell population.

Feeder cells are necessary to help single hybrid cells to develop into colonies and the tissue culture conditions, in general, should be good.

*b. Materials*

- Sterile complete RPMI and HT medium and Geys hemolytic medium.
- Sterile 24-, 48-, or 96-well plates. As indicated previously (Section VI.C.4.8.), the larger wells allow more time to screen supernatants without losing the colonies through overgrowth. However, the 96-well microtiter plates are particularly suited to cloning, because a sufficient number of wells at each cell concentration is required to ensure effective cloning.
- Spleen cells from normal unimmunized mice (one spleen will provide feeder cells for at least ten cloning plates).
- Cell suspension of hybridoma cells, with viability preferably >80%.

*c. Procedure*

1. Prepare suspension of spleen cells as described in Section VI.C.3, except that the cells should finally be suspended at $2 \times 10^5$/mℓ in HT medium.
2. Prepare hybridoma cells in HT medium, at concentrations of 100, 30, and 10 cells/mℓ. Note that in order to achieve dilutions of this order, it is better to make several dilution steps rather than taking a very small volume and diluting to a large volume. Thus, if the cells are at $5 \times 10^5$/mℓ, take 100 μℓ and dilute to 10 mℓ (to give $5 \times 10^3$/mℓ); take 100 μℓ of this and dilute again to 5 mℓ, giving 100 cells/mℓ, and make final dilutions from this stock. At each stage ensure that the cells are evenly mixed before sampling.
3. Mix feeder and hybridoma cells in equal volumes, giving a uniform feeder cell concentration of $0.5 \times 10^5$/mℓ and hybridoma cell concentrations of 50, 15, and 5/mℓ. Aliquot 200 μℓ amounts, using one 96-well plate per hybrid, and four columns at each cell concentration. Alternatively, prepare feeder layers by plating out 100 μℓ aliquots of cell suspension at $10^5$/mℓ and later add 100 μℓ aliquots of the hybridoma cells at 100, 30, and 10 cells/mℓ. Feeder layers prepared up to 7 days in advance will still support cloning. This is convenient when hybrids are ready for cloning at different times.

4. Observe every 2 to 3 days for the next 10 to 14 days, during which colonies should appear. There is no need to feed until wells are showing growth and yellowing, but it is important to note wells which, under the microscope, appear to have multiple colonies, or colonies which are markedly asymetrical and may derive from two adjacent cells. Such wells should be excluded from further consideration. Other wells are tested to ensure they are secreting the desired antibody, as identified by the screening assay. Positive colonies should be expanded gradually, through 2- and 10-mℓ cultures, and cryopreserved.

*2. Semisolid Agar Cloning*

This technique also requires dilution of the cells so that single colonies grow, but because the colonies are grown in a semisolid medium, cells are not free to move. It has been argued that this technique gives a greater certainty of monoclonality. However, the main reason in either technique for uncertainty is that a pair of different cells may lie adjacent and give rise to a single, mixed colony. This is no less likely in the agar technique. Indeed, cloning efficiencies are much lower in agar, necessitating the use of higher cell concentrations (1000 to 10,000 cells per milliliter as compared to 1 to 10 cells/well in the limiting dilution method). This increases the chances of colonies arising from adjacent cells in the agar technique. Furthermore, the semisolid medium procedure is significantly more complicated than the liquid medium method. The semisolid agar method is used relatively infrequently and will not be described here. A detailed description may be found in Campbell.[34]

*3. Cloning and Selection by Flow Sorting*[35]

Fluorescence-activated cell sorters can be fitted with single-cell deposition systems, which place a single cell with the selected properties in a culture well and then cut off. The culture plate is then moved manually to the next well and the sorter activated again. In this way single cells can be placed into microtiter wells aseptically. The method provides no advantage over manual cloning unless the cells are selected for antigen binding. For hybridomas against soluble antigen, the purified antigen may be fluorescein labeled and mixed with the hybridoma culture. Cells bearing membrane antibody against the antigen will react with it and can be positively selected for cloning. The method appears to work well, although plasma cells have traditionally been regarded as cells which secrete antibody, but do not express it on their surface membrane.

## X. CRYOPRESERVATION

### A. Introduction and Rationale

One of the principal advantages of hybridomas, that they provide a potentially permanent source of antibody, can only be realized if cells can be preserved indefinitely. This is done by storing cells at very low temperatures — cryopreservation. Although there may be a gradual loss of viability at liquid nitrogen temperatures, cell lines have been preserved for years and then successfully reestablished in culture. The temperature in the liquid nitrogen is −196°C, while that of the vapor phase over liquid nitrogen is −156°C. In practice, ampules are in the liquid when the tank is relatively full and in the vapor phase when the level of liquid nitrogen is relatively low. Storage at higher temperatures, such as −80°C, is only suitable for short periods, days or at most weeks.

Because of the constant risk of contamination or cell death in culture, it is important to cryopreserve as early as possible. Five to ten vials should be stored for each cell line, and the stock should be maintained at a minimum of five vials, by storing more from a reestablished culture. An adequate bookkeeping system is required, to ensure that the last vial is not thawed without this fact becoming apparent. It is important, particularly before a laboratory has built up extensive experience, to check that cultures can be reestablished from specimen-stored vials. A fail-safe system for keeping the cell bank supplied with liquid nitrogen is essential, ensuring that holidays and sickness are planned for. Liquid nitrogen supplies can be interrupted, and it is a good idea to deposit vials of cells for storage in another institution.

The process of cooling the cells to liquid nitrogen temperatures and the process of thawing to reestablish cultures are both potentially damaging to cells and must be done carefully. There are a variety of procedures which work equally well; the procedures routinely used in the author's laboratory are set out below. The essential features of the methods involve reducing the amount of damage caused by the formation of crystals, by the addition of dimethylsulfoxide (DMSO), and by cooling the cells according to a program designed to reduce the size of ice crystals.

## B. Freezing Cells Down

1. Cells are taken from a healthy culture (viability > 80%); split to $3 \times 10^5$ the previous day), centrifuged, and resuspended at $6 \times 10^6$/mℓ in HT medium containing 50% FCS; 1 mℓ of cell suspension is needed for each vial to be stored.
2. To this suspension, an equal volume of RPMI medium (without serum) containing 30% DMSO is added dropwise, with continuous gentle mixing, at a rate of 1 mℓ/min.
3. After gentle but thorough mixing the cell suspension is dispensed in 2 mℓ aliquots in freezing vials. These may be sterile glass vials which must be sealed with a flame or, more conveniently, special plastic freezing vials with screw-on caps. Vials with push-on caps should not be used, since liquid nitrogen can leak into these and vaporize explosively during subsequent thawing.
4. After ensuring that the vials are sealed, they are transferred to a programmed cooling instrument, if available. These machines admit liquid nitrogen to the cooling chamber at a rate controlled by a cam. The shape of the cam controls the rate of cooling and produces different rates at different stages of the program. Various designs of cam are available, and there is a lack of detailed study on the optimal program for cooling. However, cams supplied with cooling instruments work satisfactorily and are generally designed to reduce the temperature from ambient to −10°C at 1 to 2°C/min, increase the rate of admission of nitrogen at −10°C to take up the latent heat of fusion as the mixture freezes, continue cooling at 1 to 2°C/min to −25°C, and then increase the rate gradually to 5 to 10°C/min. The cam stops when the temperature reaches −100°C, at which time the vials should be transferred rapidly and with minimal handling to the vapor phase of the liquid nitrogen container.

If a programmed cooling device is not available adequate preservation of cells can be achieved by placing the vials initially in a container made of polystyrene at least 1 cm thick, and placing this container in the top of the liquid nitrogen container or a −80°C freezer. The rate of cooling is low because of the insulation provided by the polystyrene. The cells are left overnight in the polystyrene container and then transferred to the body of the tank.

### *1. Note*

Once the DMSO has been added, the process should be completed as rapidly as possible, without interruption, since DMSO is toxic to cells in the liquid phase.

### C. Thawing Cells Out

1. Remove vial from liquid nitrogen container and place immediately in a 37°C water bath until ice has just melted; do not leave in water bath longer than is necessary to just melt.
2. Dilute cell suspension by slow dropwise addition of an equal volume of HT medium at room temperature. The medium (2 mℓ) is added over a period of 10 min, with gentle mixing. When the volume is too large for the freezing vial the contents are gently transferred to a 10-mℓ tube to continue dilution.
3. The cell suspension is allowed to stand at room temperature for 15 min, and then further diluted with 6 mℓ HT medium added dropwise over 10 min.
4. The cell suspension is again left to stand for 15 min.
5. The cells are centrifuged at 200 g for 5 min, resuspended in 10 mℓ HT medium, and placed in a pregassed culture flask. At this stage it is advisable to check the viability. If it is less than 50%, the chances of recovering viable cultures can be improved by placing the cells on a feeder layer, prepared as described for cloning (Section IX.B.1). The contents of the ampule may be placed in a well of a 6-well cluster dish, to which feeder cells have been added previously. Alternatively, the thawed cells are mixed with an equal volume of feeder cells at $2 \times 10^6$/mℓ to give a total of 10 mℓ, and placed in a culture flask or well. When cell growth is apparent cells may be transferred to culture vessels without feeder cells.

#### *1. Notes*

1. We thaw in HT medium because the cells are grown in this medium.
2. A variety of alternative procedures have been described, including the much quicker method of thawing out the cells and diluting them immediately in 10 mℓ medium. While the fast dilution method works, the viability of the cells immediately after thawing is significantly reduced, as compared with the slow dilution method described above. Thus, the fast dilution method is not recommended unless the viability of the stored cells is good.

## XI. PROBLEM SOLVING

It is a common experience to find that after successfully initiating hybridoma work the laboratory goes through a "bad patch" during which few, if any, colonies are produced. Clearly, there are many potential sources of trouble and an even larger number of solutions. One approach which is of proven practical value when fusions are not going well is to test the individual components of the media in cloning assays, with an established hybridoma line. Different batches of the basic medium (including a batch of high-quality medium bought ready-made) can be compared; different batches of serum, antibiotics, and other additives and feeder cells can be used in conjunction with a commercial batch of basic medium. This approach will not necessarily solve the problem, but in the author's experience it has done so on two occasions. In both instances medium made up from commercially supplied powders were at fault, and this was made obvious by comparison with medium supplied as liquid. In one case the medium had not been made up correctly; in the other it had not been sterilized immediately, leading probably to the presence of pyrogen in the final medium.

Other reasons for the failure of cultures to thrive include fluctuations in temperature, humidity, and $CO_2$ concentration in the incubator. As discussed previously, hybrids should not be taken off HAT and put straight into normal medium, because the aminopterin persists and will kill cells in the absence of added hypoxanthine and thymidine. Cultures are fed with HT medium, until all traces of aminopterin have disappeared. In practice, we (and many other workers) always feed hybridomas with HT rather than wean them off it.

Occasional contamination of cultures is almost inevitable, but frequent contamination must be investigated. Sterility is discussed in Chapter 4, Section II.A.3. Identification of contaminating organisms can provide clues as to their origin. Fungal spores frequently come from the skin of the tissue-culture worker, but a high frequency of fungal-contaminated cultures indicates that the incubator needs to be disinfected. The laminar flow hood should be checked at regular intervals (specified by the manufacturer), and sterilization procedures must be carried out correctly. Autoclaving (as opposed to dry heat sterilization) relies on the latent heat given out as steam condenses to kill microorganisms. Thus, equipment must be wet or the steam must be able to penetrate all parts of the apparatus. This applies to filtration membranes, which should be moistened before autoclaving or left unsealed so that steam can have access to the membrane and plumbing. Waterbaths support microbial growth; disinfectants such as "Panabath" from BDH may be added to prevent this. A bottle of medium which has been in the waterbath should be washed down with alcohol before being introduced into the laminar flow hood. Animals should be kept out of the laboratory, except when it is necessary to remove a spleen for fusion or for feeder layer production, and the animal should be thoroughly soaked with alcohol before introduction into the sterile area. When tissue culture and animal work have to be done on the same day, we prefer to do the animal work last, to reduce the chances of bringing microorganisms from the animal house into the culture area.

Cultures which go yellow without much growth of the mammalian cells and without any detectable contaminant may be infected with mycoplasma. Mycoplasma-infected cultures can struggle along without major breakdown in growth for years, and mycoplasma detection tests have a significant failure rate. As a result, laboratories may carry infected cultures without knowing, and poor practices can lead to cross-infection. A brief discussion of mycoplasma and a reference for further information may be found in Chapter 4, Section II.A.3.

## REFERENCES


1. Westerwoudt, R. J., Improved fusion methods. IV. Technical aspects, *J. Immunol. Methods*, 77, 181, 1985.
2. Olsson, L. and Kaplan, H. S., Human-human monoclonal antibody-producing hybridomas: technical aspects, in *Methods in Enzymology*, Vol. 92, Langone, J. J. and Van Vunakis, H., Eds., Academic Press, London, 1983, 3.
3. Buck, D. W., Larrick, J. W., Raubitsbitschek, A., Truitt, K. E., Senyk, G., Wang, J., and Dyer, B. J., Production of human monoclonal antibodies, in *Monoclonal Antibodies and Functional Cell Lines: Progress and Applications*, Kennett, R. H., Bechtol, K. B., and McKearn, T. J., Eds., Plenum Press, New York, 1984, 275.
4. Mishell, B. B. and Shiigi, M., *Selected Methods in Cellular Immunology*, W. H. Freeman, San Francisco, 1980, 58.
5. Herbert, W. J., Mineral-oil adjuvants and the immunization of laboratory animals, in *Handbook of Experimental Immunology*, Vol. 3, Weir, D. M., Ed., Blackwell Scientific, Oxford, 1978, A3.1.
6. Stahli, C., Staehelin, T., Miggiano, V., Schmidt, J., and Haring, P., High frequencies of antigen-specific hybridomas: dependence on immunization parameters and prediction by spleen cell analysis, *J. Immunol. Methods*, 32, 297, 1980.
7. Trucco, M. M., Stocker, J. W., and Ceppellini, R., Monoclonal antibodies against human lymphocyte antigens, *Nature (London)*, 273, 666, 1978.

8. Zola, H., Antisera with specificity for human T lymphocytes, *Transplantation*, 24, 83, 1977.
9. Spitz, M., Spitz, L., Thorpe, R., and Eugui, E., Intrasplenic primary immunization for the production of monoclonal antibodies, *J. Immunol. Methods*, 70, 39, 1984.
10. Kohler, G. and Milstein, C., Continuous cultures of fused cells secreting antibody of predefined specificity, *Nature (London)*, 256, 495, 1975.
11. Kohler, G., Howe, C. S., and Milstein, C., Fusion between immunoglobulin secreting and nonsecreting lines, *Eur. J. Immunol.*, 6, 292, 1976.
12. Kearney, J. F., Radbruch, A., Liesengang, B., and Rajewsky, K., A new mouse myeloma cell line that has lost immunoglobulin expression but permits the construction of antibody-secreting hybrid cell lines, *J. Immunol.*, 123, 1548, 1979.
13. Schulman, M., Wilde, C. D., and Kohler, G., A better cell line for making hybridomas secreting specific antibodies, *Nature (London)*, 276, 269, 1978.
14. Fazekas, de St. Groth S. and Scheidegger, D., Production of monoclonal antibodies: strategy and tactics, *J. Immunol. Methods*, 35, 1, 1980.
15. Trowbridge, I. S., Interspecies spleen-myeloma hybrid producing monoclonal antibodies against mouse lymphocyte surface glycoprotein T200, *J. Exp. Med.*, 148, 313, 1978.
16. Galfre, G., Milstein, C., and Wright, B., Rat × rat hybrid myelomas and a monoclonal anti-Fd portion of mouse IgG, *Nature (London)*, 277, 131, 1979.
17. Lachmann, P. J., Olroyd, R. G., Milstein, C., and Wright, B. W., Three rat monoclonal antibodies to human C3, *Immunology*, 41, 503, 1980.
18. Bazin, H., Production of rat monoclonal antibodies with the LOU rat non-secreting IR983F myeloma cell line, in *Protides of the Biological Fluids 29th Colloquium*, Peeters, H., Ed., Pergamon Press, New York, 1981, 615.
19. Littlefield, J. W., Selection of hybrids from matings of fibroblasts in vitro and their presumed recombinants, *Science*, 145, 709, 1964.
20. Kenny, P. A., McCaskill, A. C., and Boyle, W., Enrichment and expansion of specific antibody-forming cells by adoptive transfer and clustering, and their use in hybridoma production, *Aust. J. Exp. Biol. Med. Sci.*, 59, 427, 1981.
21. Bankert, R. B., DesSoye, D., and Powers, L., Antigen-promoted cell fusion: antigen-coated myeloma cells fuse with antigen-reactive spleen cells, *Transpl. Proc.*, 12, 443, 1980.
22. Lo, M. M., Yow Tsong, T., Conrad, M. K., Strittmatter, S. M., Hester, L. D., and Snyder, S. H., Monoclonal antibody production by receptor-mediated electrically induced cell fusion, *Nature (London)*, 310, 1984.
23. Galfre, G., Howe, S. C., Milstein, C., Butcher, G. W., and Howard, J. C., Antibodies to major histocompatibility antigens produced by hybrid cell lines, *Nature (London)*, 266, 550, 1977.
24. Gefter, M. L., Margulies, D. H., and Scharff, M. D., A simple method for polyethylene glycol-promoted hybridization of mouse myeloma cells, *Somatic Cell Genet.*, 3, 231, 1977.
25. Oi, V. T. and Herzenberg, L. A., Immunoglobulin-producing hybrid cell lines, in *Selected Methods in Cellular Immunology*, Mishell, B. B. and Shiigi, S. M., Eds., W. H. Freeman, San Francisco, 1980, 351.
26. Brodsky, F. M., Parham, P., Barnstable, C. J., Crumpton, M. J. and Bodmer, W. F., Monoclonal antibodies for analysis of the HLA system, *Immunol. Rev.*, 47, 3, 1979.
27. Astaldi, G. C. B., Janssen, M. C., Lansdorp, P., Willems, C., Zeijlemaker, W. P., and Oosterhof, F., Human endothelial culture supernatant (HECS): a growth factor for hybridomas, *J. Immunol.*, 125, 1411, 1980.
28. Zola, H., Gardner, I., Hohmann, A., and Bradley, J., Analytical flow cytometry in the study of hybridization and hybridomas, in *Proc. 6th Australian Biotechnology Conf.*, Doelle, H. W., Ed., University of Queensland, Brisbane, Australia, 1984, 405.
29. Mason, D. Y., Cordell, J. L., and Pulford, K. A. F., Production of monoclonal antibodies for immunocytochemical use, in *Techniques in Immunocytochemistry*, Vol. 2, Bullock, G. R. and Petrusz, P., Eds., Academic Press, London, 1983, 175.
30. O'Neill, H. C. and Parish, C. R., A rapid, automated colorimetric assay for measuring antibody binding to cell surface antigens, *J. Immunol. Methods*, 64, 257, 1983.
31. Versteegen, R. J. and Clark, C., The use of streptavidin in the detection of monoclonal antibodies, in *Monoclonal Antibodies and Functional Cell Lines: Progress and Applications*, Kennett, R. H., Bechtol, K. B., and McKearn, T. J., Eds., Plenum Press, New York, 1984, 393.
32. Main, E. K., Hart, M. K., and Wilson, D. B., Immunofiltration. A rapid screening assay for detection of antibodies directed against cell surface antigens, in *Monoclonal Antibodies and Functional Cell Lines: Progress and Applications*, Kennett, R. H., Bechtol, K. B., and McKearn, T. J., Eds., Plenum Press, New York, 1984, 376.
33. Harris, H., Monoclonal antibodies to enzymes, in *Monoclonal Antibodies and Functional Cell Lines: Progress and Applications*, Kennett, R. H., Bechtol, K. B., and McKearn, T. J., Eds., Plenum Press, New York, 1984, 33.

34. Campbell, A. M., *Laboratory Techniques in Biochemistry and Molecular Biology, Monoclonal Antibody Technology,* Burdon, R. H. and van Knippenberg, P. H., Eds., Elsevier, Amsterdam, 1984.
35. Parks, D. R., Bryan, V. M., Oi, V. T., and Herzenberg, L. A., Antigen-specific identification and cloning of hybridomas with a fluorescence-activated cell sorter, *Proc. Natl. Acad. Sci. U.S.A.,* 76, 1962, 1979.

## FURTHER READING

### General Cell Culture

Freshney, R. I., *Culture of Animal Cells. A Manual of Basic Technique,* Alan R. Liss, New York, 1983.
Adams, R. L. P., *Cell Culture for Biochemists, Laboratory Techniques in Biochemistry and Molecular Biology,* Burdon, R. H. and von Kippenberg, P. H., Eds., Elsevier/North-Holland, Amsterdam, 1980.
Colowick, S. P. and Kaplan, N. O., *Methods in Enzymology,* Vol. 58, Jakoby, W. B. and Pastan, I. H., Eds., Academic Press, London, 1979.

### Monoclonal Antibody Production

Campbell, A. M., *Monoclonal Antibody Technology, Laboratory Techniques in Biochemistry and Molecular Biology,* Burdon, R. H. and van Knippenberg, P. H., Eds., Elsevier, Amsterdam, 1984.
Goding, J. W., *Monoclonal Antibodies: Principles and Practice,* Academic Press, London, 1983.
Colowick, S. P. and Kaplan, N. O., *Methods in Enzymology,* Vol. 92, Part E, Langone, J. J. and Van Vunakis, H., Eds., Academic Press, London, 1983.

Chapter 4

# MONOCLONAL ANTIBODIES: PRODUCTION, PURIFICATION, ANALYSIS, QUALITY CONTROL, STORAGE, AND DISTRIBUTION

## I. INTRODUCTION

This chapter deals with the preparation and processing of monoclonal antibodies once the hybridoma line is available. Techniques for the purification and analysis of monoclonal antibodies are described. The successful use of these procedures requires some familiarity with the techniques and equipment used in protein chemistry. Some suggestions for further reading are included for the reader who is not familiar with protein chemistry.

## II. PRODUCTION

There are two distinct ways of making monoclonal antibody from the hybridoma line: by growing the line in suspension culture and by growing it in animals as a tumor. Cell culture material is not contaminated by other antibodies but can be produced only at low concentration, while material produced in animals may be contaminated with other immunoglobulins, but can be produced in much higher titer. In most laboratories it is necessary to produce both types of material, since each has its particular uses. Tissue culture supernatant can be used for most serology, while ascites is used for immunochemical characterization or isolation of antigens. Commercial production tends to be based on ascitic fluid, because this is the cheaper way of making large amounts of antibody. In very approximate terms, ascites contains antibody 1000 times the concentration in culture supernate. Since one mouse can produce 3 to 5 mℓ ascites (or more, see below), this is equivalent to 3 to 5 ℓ of tissue culture supernatant.

### A. Tissue Culture Production

#### *1. Scale*

Antibody may be produced simply by maintaining the hybridoma in culture and "splitting" the cultures every 2 to 3 days, when the cell concentration reaches $10^6$/mℓ. Even a 10-mℓ culture will produce adequate amounts of antibody for some purposes, since many procedures can be carried out with a few microliters of antibody at the concentration secreted by cell cultures. The cultures can readily be scaled up to 50 to 100 mℓ without specialized equipment, just by using larger flasks. Standard flask sizes (expressed in surface area) are 25, 80, and 150 $cm^2$. To allow adequate gas exchange the flasks are not filled with culture medium; suitable culture volumes for the three flask sizes quoted are 10, 40, and 80 mℓ. The flasks should have their lids loosened when in the gassed incubator, to allow gas exchange. Larger volumes can be produced in simple stirred culture vessels, and specialized equipment is available for larger-scale production.

For most purposes, if culture in flasks cannot produce enough material it is best to turn to production in animals, and there is, thus, little scope for the specialized equipment required for mass culture. Applications which might merit the bulk production of antibody in culture would include the preparation of material for in vivo human use, where there may be anxiety about the contamination of antibody with viral genomic material from the animal. Anxiety about viral genomic material produced by the hybridoma would, perhaps, be more appropriate. The production of human monoclonal antibodies in bulk may also require large-scale culture, although they can be produced as ascitic fluid in immunosuppressed mice (see below).

Large-scale growth of hybridomas can be carried out in equipment designed for the large-scale suspension culture of other mammalian cell lines. Techniques for scaling up are described in an article by Acton et al.,[1] which gives references to other articles from the same group describing equipment for cultures up to 200 ℓ. An alternative to the use of larger vessels is provided by the "Vitafiber II" hollow fiber cell culture system manufactured by Amicon (U.S.). In this apparatus the medium is continuously circulated past a semipermeable membrane which encloses the cells. Low molecular-weight components of the medium can diffuse across the membrane, and low molecular-weight waste products are removed. The cells are, thus, growing effectively in medium free of serum proteins, but have access to low molecular-weight growth factors. The secreted antibody is collected periodically. The method appears well suited to the production of large amounts of antibody, particularly when production in animals is not desirable. At the time of writing, however, there are few published reports on the performance of the apparatus in practice. An alternative way of making large amounts of antibody is to contract the production to a biotechnology company which has the necessary facilities. In particular, Damon Biotech, 119 Fourth Avenue, Needham Heights, Mass. 02194 specializes in large-scale production of monoclonal antibody under conditions which are suitable for antibody which is to be used in patients.

*2. Medium*

The standard, serum-supplemented media (Chapter 3, IV.A) may be used if the volume to be produced is not large. Costs can be reduced by using less expensive sera, since established hybridoma lines do not absolutely need 10% FCS. Various sera, including mixtures of horse and FCS, adult bovine serum, newborn calf serum, and calf serum at various concentrations, may be used. For short periods (1 to 2 weeks) the serum concentration can simply be reduced to 5%. Serum-free medium may be preferred either in order to produce purer antibody or to reduce the cost of production. A number of serum-free media has been used for the growth of hybridomas. Some are available commercially, but while their composition is kept secret, the advantage over the use of serum must be considered marginal — they cannot be described as "defined media". Published recipes for serum-free media for the growth of murine hybridomas include the medium of Murakami et al.,[2] which is supplemented with insulin, transferrin, ethanolamine, and selenite ion.

*3. Sterility*

Microbial contamination is a constant threat with all tissue culture work. The media which support the growth of mammalian cells also support the growth of bacteria and fungi. Good tissue culture technique, the use of tissue culture cabinets with a flow of sterile air, and the incorporation of antibiotics in the medium reduce contamination to an acceptable frequency, but contamination will, nevertheless, occur. Three additional techniques will help manage contamination: sterility testing of media, identification of contaminants, and the use of special antibiotics in emergency situations.

*a. Sterility Testing*

The priority in the hybridoma laboratory is the production and characterization of monoclonal antibodies, and no one wants to spend time on unnecessary peripheral studies. However, experience shows that a certain level of quality control on media is more than repaid by avoiding long delays due to sterility failures. Constituents which are bought from reputable manufacturers, including FCS, can be taken as sterile, since the quality-control program of the manufacturer is much more extensive than that carried out by most hybridoma laboratories. On the other hand, constituents made up and sterilized in the laboratory should be sterility tested. Sterility testing of media is discussed in detail by McGarrity.[3] Most laboratories will be reluctant to invest the time required to carry out the full test protocol. In the author's laboratory testing is carried out for fungi (using Sabouraud medium, Code #CM147, Oxoid) and aerobic bacteria, using trypticase/soy broth. Anaerobic organisms are not tested for, since they are not generally a source of contamination in aerobic cell cultures. Our media batches normally consist of 20 ℓ, filtered and dispensed into 500-mℓ bottles. Every tenth bottle is tested. From each bottle, 100 mℓ is taken for fungal testing and 100 mℓ for aerobic bacteria. The medium is passed through a vented 0.22-$\mu$m filter designed to collect the microorganisms and spores (IVEX-HP filter assembly, Millipore Corp.). The filter is first washed through with sterile water, followed by the medium under test. Excess medium is removed by washing through with sterile water (three times), and the filter assembly is filled with the appropriate microbiological growth medium, stoppered, and incubated to allow microbial growth. The fungal test is incubated at 32°C for 14 days, while the aerobic cultures are incubated for 7 days at 23°C.

Obviously, the entire operation must be carried out aseptically, and this may be verified by preparing additional test filters which have not been challenged with the medium under test. If these show contamination, then either the operator's sterile technique or one of the other components (water, microbiological media, syringes, etc.) is at fault.

*b. Identification of Contaminants*

The identity of a contaminant will often provide a clue as to its origin. Yeast contamination is likely to derive from the skin of the person handling the cultures and suggests a need for a more careful technique and for the wearing of gloves. Bovine species of mycoplasma are likely to originate from the FCS, whereas certain bacteria are waterborne and their presence suggests inadequate filtration or the introduction of bacteria from the waterbath used to warm up the medium.

Contamination with fungi can usually be recognized by examining the culture with the inverted microscope. Low-level contamination with bacteria is less easy to identify in the inverted microscope, since there are often numbers of small cell fragments, exhibiting Brownian motion, which are difficult to distinguish visually from bacteria. A culture which is suspected of being contaminated with bacteria should be stained, using the Gram stain. This staining procedure is in routine use in most microbiology laboratories. Further characterization of the contaminant is best carried out by a specialist microbiologist. Procedures for the detection of mycoplasma and viruses are also specialized and are not usually carried out in a hybridoma laboratory. Mycoplasma testing is described by McGarrity.[3] Monoclonal antibodies being prepared for therapeutic use in patients will need to undergo more rigorous screening for viral contaminants, requiring specialized advice and assistance.

*c. Additional Antibiotics*

The best thing to do with a contaminated culture is to cap it tightly and throw it away, preferably via an autoclave. However, there are occasions when "curing" a contaminated culture is worth trying — when an important hybridoma is contaminated and there are no cells in liquid nitrogen, or when the culture stored in liquid nitrogen turns out to have a contaminant. If time allows, antibiotic sensitivity testing should be carried out on the contaminant. It should not be forgotten that most antibiotics have adverse effects on the rate of growth of the hybridomas, and that many antibiotics will inhibit, but not destroy, the contaminant.

Gentamycin, 0.2 mg/mℓ, may be used routinely in place of penicillin and streptomycin, or as an emergency replacement if a contaminant grows in the presence of penicillin and streptomycin. The routine use of gentamycin may be particularly advantageous in that it has some activity against mycoplasmas. Other antibiotics which may be useful against resistant organisms are, for Gram-positive organisms, tylosin tartrate (25 μg/mℓ) or novobiocin (200 μg/mℓ); for Gram-negative organisms, neomycin (100 μg/mℓ) or kanamycin (500 μg/mℓ). The latter is also useful against mycoplasma. Eradication of fungal contamination may be attempted by adding amphotericin B (fungizone) at 5 to 10 μg/mℓ or mycostatin at 100 μg/mℓ. Amphotericin B is supplied as a suspension and cannot be filter sterilized without loss of activity. A more complete listing and discussion of antibiotics suitable for cell culture may be found in Perlman.[4]

B. Production in Animals

*1. Rationale*

When a hybridoma is injected into an appropriate animal, it will usually grow as a transplantable myeloma, secreting large amounts of immunoglobulin. If the myeloma is introduced as an s.c. tumor the antibody will be recoverable from serum, while if the tumor is injected i.p. it will lead to the formation of ascites, a serum-like fluid in the peritoneal cavity. In mice injected i.p., both the serum and the ascitic fluid are rich in immunoglobulin. Most of this immunoglobulin is the antibody secreted by the myeloma, but some is the immunoglobulin made by the host. This may be important, since antibody against viral or other antigens encountered by the host animal may lead to confusing reactions.

Because of the increased yield of antibody, most work is done with ascitic fluid. The serum is not usually obtained from these mice, although there is no reason, other than convenience, why it should not also be used. The amount of ascites which can be obtained from an animal is variable. In mice it is easy to obtain 3 to 5 mℓ, and it is possible to obtain up to 10 mℓ. However, leaving the mice too long in an attempt to obtain a high yield of ascitic fluid often results in death of the mice before the ascites can be collected. Moreover, the mice become distressed as the volume increases and it is ethically preferable to aim for a modest yield of up to 5 mℓ.

Ascitic fluid is most readily produced in syngeneic animals, i.e., when the myeloma parent line, the spleen cell donor, and the animal used to produce the ascitic fluid are genetically identical, at least at the major histocompatibility loci. In the vast majority of work this means using BALB/c mice both for immunization and hybridoma production, since the myelomas used most commonly for fusion derive from this strain of mouse. Rat monoclonal antibody has been produced both as serum and ascitic fluid (Kilmartin et al.[5]). If the strain or species of the hybridoma differs from the animal being used to produce ascites, it will be necessary to prevent the immunological rejection of the tumor by the host. This may be achieved by using congenitally athymic ("nude") mice or rats, or by immunosuppressing the animals as described below. Even with syngeneic animals, it is necessary to pretreat the peritoneal cavity to increase the likelihood of hybridoma growth. The material most often used is a hydrocarbon, pristane. The mechanism of action is not known, but pristane probably acts as an irritant rather than as an immunosuppressive agent. If a hybridoma regularly fails to produce ascites under the standard protocol it may be necessary to immunosuppress the recipient, even though the mice and hybridoma are thought to be syngeneic.

Animals may be immunosuppressed with antilymphocyte serum, immunosuppressive drugs or radiation, or combinations of these procedures. We have used 0.5 mℓ of rabbit-antimouse or horse-antimouse lymphocyte serum, prepared by well-established methods.[6] A sublethal dose of radiation (600 rad) is as effective, but some animals may succumb to infection. Weissman et al.[7] have used a combination of hydrocortisone and radiation which permitted the production of rat and human monoclonal antibody ascitic fluid in mice.

*2. Procedure: Production of Ascitic Fluid in BALB/c Mice*

*a. Mice*

Mice, male or female, preferably 6 to 8 weeks old when first injected with pristane (see below). Brodeur et al.[8] found that male mice produced significantly more antibody and that the optimal age range was 6 to 11 weeks.

*b. Pristane*

The mice are injected with 0.5 mℓ pristane (2,6,10,14-tetramethylpentadecane) 1 to 2 weeks before injection of the hybridoma. Brodeur et al.[8] found the optimal time interval between the injection of the pristane and hybridoma to be 14 days, while Hoogenraad et al.[9] found 10 days to be optimal. Mice which have been pristane treated and then not used for a period of more than 4 weeks have been reinjected with pristane and then injected with myeloma 1 week later. These mice have produced ascites in good volume. A method for holding a mouse for i.p. injection is illustrated in Figure 3, Chapter 3.

*c. Hybridoma*

The optimal number of cells required to produce growth appears to vary from hybrid to hybrid. After experimenting with different doses of cells we have standardized on the injection of $2.5 \times 10^6$ cells, although $10^6$ cells will generally produce ascites. Brodeur et al.[8] demonstrated a broad optimum dose at 6 to $32 \times 10^5$ cells per mouse, with larger doses leading to shorter survival and smaller ascitic fluid volumes. While it is unlikely that injection of an excess of cells will reduce the probability of initiating tumor growth, it is possible that the mice will die earlier because of systemic effects of the tumor.

The cells used for ascites production are generally produced in tissue culture, but if large volumes of ascites are needed cells obtained from the ascites fluid can be serially passaged through mice. Cells harvested from ascites production can also be cryopreserved to initiate later ascites production.

Some hybridomas have a tendency to produce solid tumors. Even with such "difficult" hybridomas, the occasional mouse will make ascitic fluid and the cells from this fluid may be passaged on the assumption that these cells may have been selected for growth in ascitic form. Mice which produce solid tumors may have very high concentrations of antibody in the serum.

*d. Harvesting*

Growth may take 7 to 21 days and, as indicated previously, it is advisable, and less distressing to the animals, to aim for a rather modest yield of about 5 mℓ. The degree of swelling which will give this yield will be learned by experience, but as a rough guide the mouse looks pregnant. Mice must be checked frequently, at least three times a week and daily in the later stages of tumor growth. The mouse is killed by $CO_2$ or cervical dislocation: the abdominal skin swabbed with disinfectant, and a small incision made in the abdomen, over the peritoneum and slightly to one side. The skin is pulled away, and the peritoneal cavity wall is then pinched up with a pair of tweezers and stretched slightly to make a pouch into which the Pasteur pipette will be inserted to collect the fluid (Figure 1). With the pipette ready, an incision is made and, still holding the peritoneal wall with the tweezers, the fluid is collected. There is a risk at this stage of the fluid leaking out if the tweezers are not used properly. The procedure is illustrated

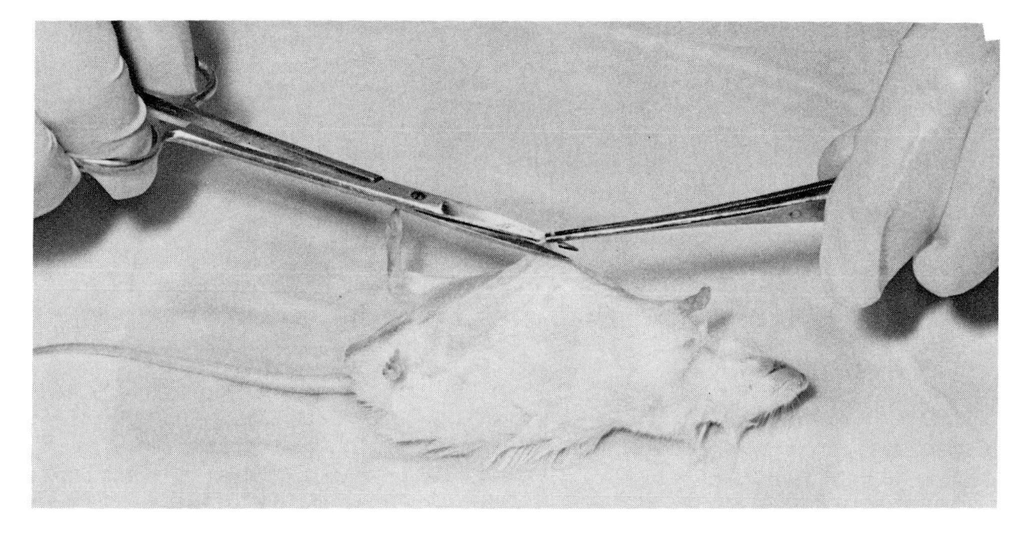

A

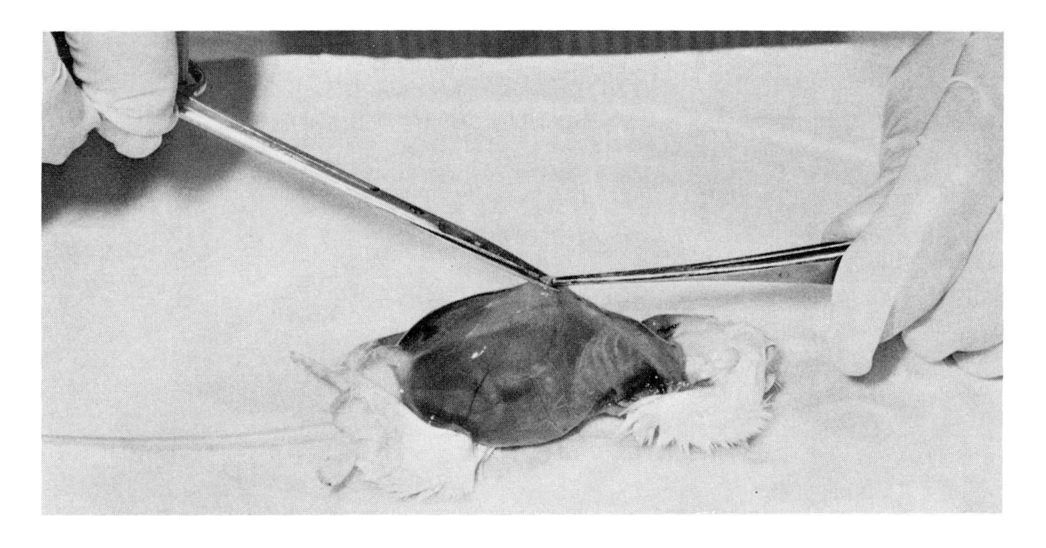

B

FIGURE 1. Harvesting of ascites. (A) Make an incision in the skin; (B) pull the skin well out of the way and, using sterile instruments, make an incision in the wall of the peritoneal cavity; (C) withdraw ascitic fluid with a Pasteur pipette, creating a "pocket" for the fluid to collect in once the bulk of the fluid has been taken.

in Figure 1. If harvesting and processing are carried out using aseptic technique, the ascites will be sterile. This will facilitate later use of the material, since sterilizing small volumes of liquid is a wasteful process.

*e. Processing of Ascites*

If the cells are to be used, the ascites is centrifuged gently (200 g for 5 min) and the cells separated from the supernatant. If the cells are not required, this preliminary step may be omitted. The ascites must be clarified and contains fat as well as cells. Centrifugation at 10,000 × G for 15 min separates the ascites into a sediment, clear superna-

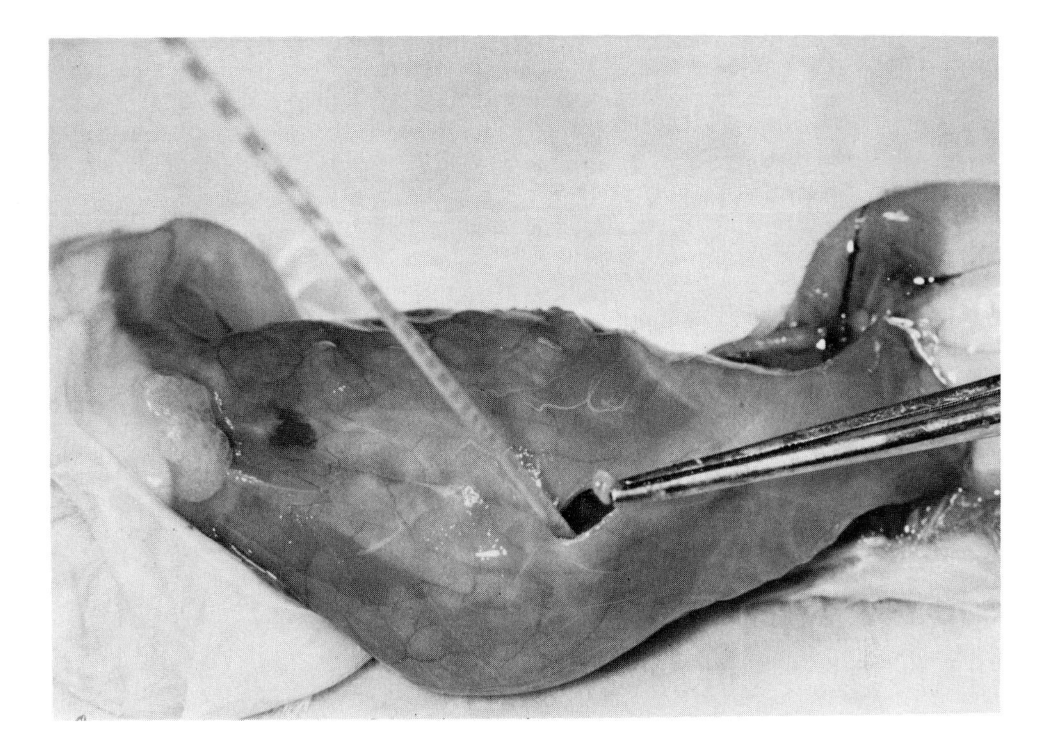

FIGURE 1C

tant, and plug of fat at the top of the supernatant layer. The supernatant may be obtained fairly free of fat by carefully piercing the fatty layer with a Pasteur pipette and drawing off the supernatant. If the tube is tilted the fatty layer will adhere to the sides of the tube as the liquid is removed from underneath. Some material is sacrificed in order to obtain clear liquid, and some fat may be disturbed and contaminate the liquid. This can be removed later. An alternative procedure for defatting involves the addition of silicon dioxide, Cab-O-Sil (Union Carbide), which aggregates the fat.[10] The silicon dioxide powder must be handled with care, since if inhaled it can cause lung damage. Ascitic fluid is diluted with an equal volume of buffer (the choice of buffer depends on the purification procedure to be used subsequently, see Section III). To 20 mℓ of diluted fluid add 0.28 g Cab-O-Sil powder, mix for 30 min at room temperature, and centrifuge. The supernatant is clear and can be directly filtered through a sterilizing membrane. Ascitic fluid sometimes produces a clot, which can be allowed to retract and can be removed in the same way that serum is obtained from clotted blood.

It is advisable to heat inactivate the ascites at this stage, in order to inactivate the components of complement, which may interfere in later tests, and to destroy some proteolytic enzymes which may degrade the antibody. Heat inactivation is carried out by treating the ascitic fluid, in a glass vial (plastic vials tend to release extraneous chemicals into the fluid on heating) at 56°C for 30 min. These conditions should not be varied. A procedure used to make aggregates of IgG involves heating at 60°C, indicating how little leeway there is in varying the conditions without damaging the antibody. Some monoclonal antibodies lose significant amounts of activity on heat inactivation. Underwood and Bean[11] surveyed 38 monoclonal antibodies and found 1/8 IgM and none of the IgG antibodies lost activity on treatment at 56°C for 30 min. Our own experience, however, is that loss of activity is common amongst IgM antibodies and does occur with some IgG antibodies. Thus, it is important to check the stability of each particular monoclonal antibody under the conditions of heat inactivation. This property would not be expected to vary from batch to batch and the experiment needs to be done only once for each clone. If the antibody is not heat inactivated it is advisable to purify it in order to remove complement components and proteolytic enzymes. A convenient way of testing for proteases and confirming their removal is provided by the Protease Substrate Gel tablets (Bio-Rad Laboratories).

The ascites may now be tested, stored, or used. It may be advisable to titrate ascites from individual mice before pooling, in case some mice have made inactive material. Each new pool should be titrated and, if the monoclonal immunoglobulin is to be purified, the protein electrophoresis pattern should be established (Section III.A (number 1 in the list) and Section IV.B.3) to ensure that there is a good monoclonal band.

#### *f. Storage*

This is discussed in greater detail later (Section V.A). At this stage the ascites can be stored at −20 or −80°C, as a bulk with two to three small (100 μℓ) samples for testing.

## III. PURIFICATION

### A. Rationale

For the majority of applications, there is no need to purify the immunoglobulin from either culture supernatant or ascitic fluid. If purification is required, the method should be selected to match the degree of purity needed. If pure monoclonal antibody is essential, ascites is not a suitable starting material, because it contains immunoglobulin derived from the immune system of the ascites-producing animal. Otherwise, ascites is the preferred starting material because of the larger amount of antibody available. For many purposes it is adequate to purify the total immunoglobulin fraction from ascites. It is possible to enrich selectively for the hybridoma protein by selecting the fraction on the basis of charge. The hybridoma protein will, nevertheless, be contaminated with some normal immunoglobulin.

Many methods have been used to isolate immunoglobulin from sera, but some of them do not work well with mouse immunoglobulin. The two methods used most commonly to purify IgG monoclonal antibody depend, respectively, on binding to staphylococcal protein A and anion exchange chromatography. The methods have different merits and disadvantages, and both are described. The protein A method is less suited to large-scale work because of the cost of the affinity column, while ion exchange materials are relatively cheap. The principal difference, however, is that the protein A method selectively adsorbs the monoclonal protein, whereas an anion exchange column adsorbs the anionic proteins, principally albumin, and allows the immunoglobulin to pass through unretarded. The anion exchange method leaves cationic proteins, such as transferrin, contaminating the immunoglobulin. The optimal size of protein A column to be used depends on the amount of immunoglobulin present, since the column absorbs the immunoglobulin. The optimal size of anion exchange column, on the other hand, depends on the amount of anionic protein present. Using the right size of column is important, because if the column is too large the yield suffers. An anion exchange column which is too small for the load applied will allow albumin through to contaminate the immunoglobulin. A protein A column which is too small will not absorb all the immunoglobulin from the starting material. Ascitic fluids are variable in immunoglobulin content. Ascites which contain a small amount of immunoglobulin and a large amount of albumin will require a large ion exchange column, and the yield of immunoglobulin will, consequently, suffer. Because of this, it is advisable to remove much of the albumin by a preliminary enrichment step involving precipitation of the immu-

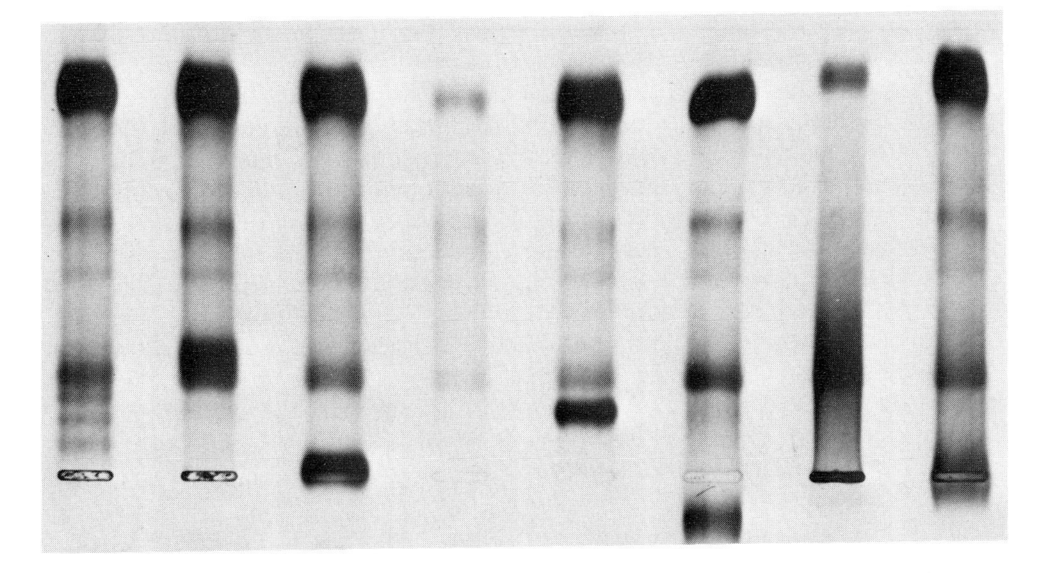


FIGURE 2. Agarose gel electrophoresis of ascitic fluids prepared from eight different hybridomas showing variation in position and intensity of the monoclonal bands. Different batches prepared from the same hybridoma should have the same monoclonal band.

noglobulin by polyethylene glycol (PEG), ammonium sulfate, or sodium sulfate. For some purposes there is no need to purify the antibody beyond such a precipitation step.

These considerations lead to the following approach to deciding which method to use:

1. Examine the material to be fractionated, determining the immunoglobulin class, protein content, and the electrophoretic profile. Figure 2 shows electrophoretic profiles obtained with several ascitic fluids, illustrating the variation in the concentration of the monoclonal band.
2. Decide on the level of purity required. If the antibody is required for coating ELISA plates or preparing immunoabsorbents the antibody does not need to be pure, but does require enrichment so that the available protein-binding sites are not occupied principally by albumin. In such cases the precipitation methods may be adequate, and the PEG method is to be recommended over the others in terms of both yield and purity.
3. For IgG antibodies use the protein A method for maximum purity. All IgG ascites with a low concentration of monoclonal immunoglobulin should be fractionated by the protein A method, if they are fractionated at all.
4. If the purity obtainable by PEG precipitation is not adequate and the immunoglobulin concentration is high, PEG-precipitated material may be further purified either by ion exchange or by gel permeation chromatography.
5. If the monoclonal antibody is IgM the protein A method is not effective, because IgM does not bind to protein A and the ion exchange method does not separate the IgM from IgG antibodies. IgM may be enriched by PEG precipitation or by gel permeation chromatography.

6. Preparative HPLC (high pressure liquid chromatography) equipment is available, at a price. Bio-Rad markets HPLC equipment designed specifically for monoclonal antibody purification, but other preparative HPLC equipment can also be used. A unique feature of the Bio-Rad system is a high-performance hydroxylapatite column with a capacity to fractionate 100 to 300 mg of antibody per cycle. This scale is well beyond the needs of most users, but will be of particular value to commercial suppliers or in projects involving large-scale clinical use.
7. Evaluation: whichever method is used, the purity and yield should be evaluated to check that the method is operating satisfactorily. Electrophoretic analysis of purity, coupled with measurement of protein, is usually adequate to assess yield, but it is important to check that the antibody activity is not lost. Titration of antibody activity, using the traditional doubling dilutions, provides only a very crude indication of recovery of activity. Most methods give titers which are reproducible to +/− one or two dilutions, encompassing a 4- to 16-fold variation in antibody concentration. However, titration will at least show if there is any gross destruction of antibody during purification.

The PEG, protein A, and ion exchange methods are described in detail below. Gel filtration is limited in capacity and is only likely to be of use for the purification of IgM. If gel filtration is used as the sole fractionation technique, the IgM will be contaminated with $\alpha$-2-macroglobulin. Gel filtration media and techniques are described in the technical literature supplied by the major manufacturers of gel filtration media (Pharmacia, LKB, and Bio-Rad). Suggestions for further reading in protein separation techniques may be found at the end of the chapter.

### B. PEG Precipitation

#### *1. Principle*

Polyethylene glycol (PEG) precipitates proteins, and the concentration of PEG required depends on the size and charge of the protein. Immunoglobulins are precipitated at concentrations which leave most other serum proteins in solution and can, thus, be purified by PEG purification. The PEG concentration which gives the best purity and yield must be determined for each monoclonal antibody by carrying out a series of precipitations on small volumes and analyzing the result.

#### *2. Materials*

1. Polyethylene glycol (PEG) MW 6000 (BDH).
2. Veronal-buffered saline (VBS) pH 7.2 made up using buffer tablets (Oxoid, Code BR 16).
3. IgM buffer: VBS containing added NaCl to 0.5 *M*, PEG stock solution at 30% wt/vol in VBS.

#### *3. Procedure*[10]

1. Clarify ascites fluid using Cab-O-Sil (Section II.B.2.e).
2. Determine optimal PEG concentration for the particular monoclonal antibody, unless already known. Take 50 $\mu\ell$ aliquots of ascitic fluid and add equal volumes of PEG solutions from 30% W/V to 5% in steps of 5%. Mix well and place tubes in ice for 15 min. Centrifuge in a refrigerated centrifuge at 2000 × G for 20 min. Collect supernatant and examine by agarose electrophoresis (Section IV.B.3). As the PEG concentration is increased the amount of monoclonal protein in the

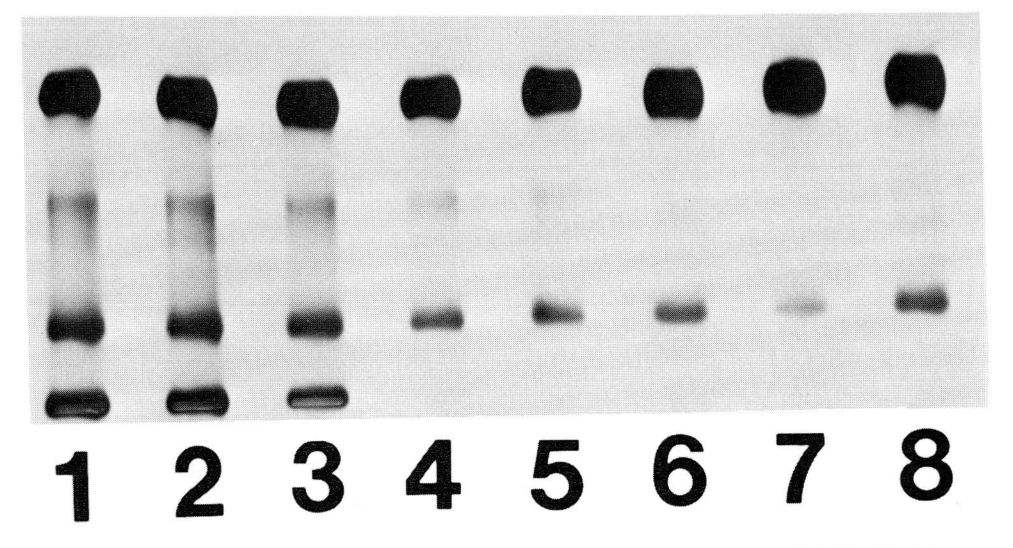


FIGURE 3. Agarose gel electrophoresis of supernatants from PEG precipitation of FMC56 ($IgG_1$) monoclonal antibody. Final PEG concentrations, from left to right: 0, 2.5, 5, 7.5, 10, 12.5, 15, and 17.5%. A PEG final concentration of 8% was selected for scaled-up fractionation of FMC56.

supernatant falls, while the concentration of the other proteins begins to fall when the PEG concentration increases above 7% (final) (Figure 3). Choose a concentration which precipitates most of the monoclonal band for the scaled-up fractionation.

3. Add 1 vol PEG (at twice the chosen final concentration) to 1 vol clarified ascites. Mix well and put tubes on ice for 15 min.
4. Centrifuge in a refrigerated centrifuge at 2000 × G for 20 min. Collect supernatant and keep for subsequent analysis.
5. Redissolve the pellet in 1 vol of veronal buffer and reprecipitate with an equal volume of PEG as before. Mix well and leave on ice for 15 min.
6. Centrifuge as above.
7. Resuspend the pellet in the buffer to the original volume. IgG can be resuspended in any buffer required for the next application of the antibody (for example, the starting buffer for ion exchange chromatography), but IgM requires a high ionic strength, such as 0.5 *M* NaCl in VBS, to redissolve the protein fully. Leave on ice for 60 min, mixing occasionally to allow the immunoglobulin to redissolve. Centrifuge again to remove any insoluble material. The supernatant is the immunoglobulin fraction.
8. Test samples: the immunoglobulin fraction should be tested by electrophoresis to determine purity. Yield and purity can be determined by scanning the electrophoretogram, to determine the % of total protein which is immunoglobulin, and measuring the total protein (Section IV.A).

*4. Notes*

The PEG method gives IgM at >90% purity if lipid is first removed using Cab-O-Sil. Because of the higher concentrations of PEG required to precipitate IgG, other proteins are precipitated as well, resulting in a lower purity of IgG (30 to 40%). This material is adequate for many purposes and can be applied to an ion exchange column (see below) if higher purity is required. Figure 4 shows the results of purification of an IgG and an IgM monoclonal antibody by the PEG method. The procedure takes about 2 hr, when the PEG concentration to be used is already known.

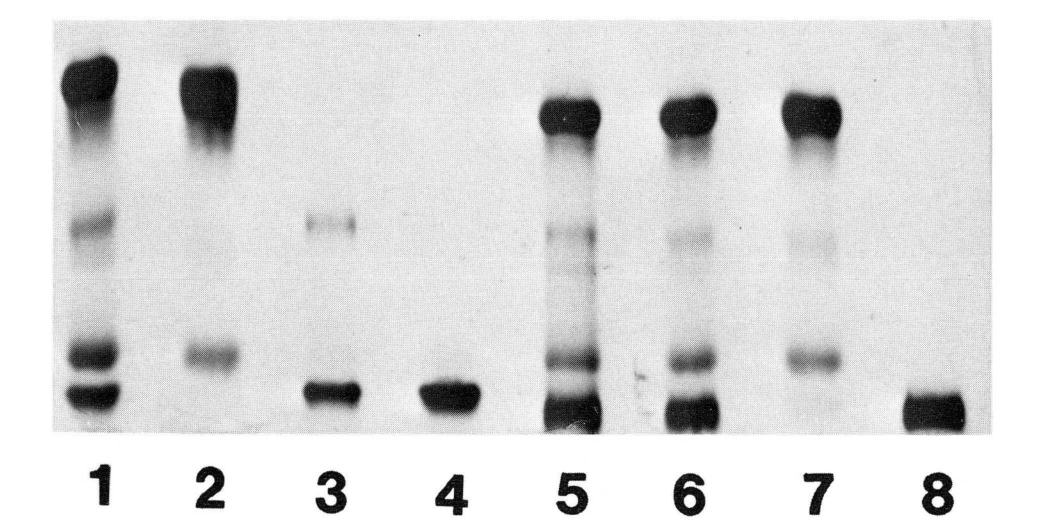


FIGURE 4. Purification of monoclonal antibodies FMC32 ($IgG_1$) and FMC24 (IgM). (Lane 1) FMC32 ascitic fluid clarified with silicon dioxide; 20 mℓ ascites, 12% monoclonal band; protein content 12.6 mg/mℓ; total monoclonal immunoglobulin 30 mg; (lane 2) supernatant of PEG precipitation (15% final PEG concentration); (lane 3) PEG precipitate; (lane 4) eluted fraction from DE52 column; 98% in monoclonal band; protein content 2.3 mg/mℓ; volume 13 mℓ; yield 97%; (lane 5) FMC24 ascitic fluid: 10 mℓ ascitic fluid, 27% monoclonal band; total protein 15.9 mg/mℓ, total starting monoclonal immunoglobulin 42.9 mg; (lane 6) FMC24 ascitic fluid after silicon dioxide treatment; (lane 7) supernatant of PEG precipitation (6% final PEG concentration); (lane 8) PEG precipitate: 95% pure monoclonal band; total protein 6.7 mg/mℓ; volume 5 mℓ; yield 74%.

## C. Ion Exchange Method

### *1. Materials*

1. Diethylaminoethyl-cellulose (Whatman DE52) column (30 mℓ disposable syringe will do).
2. Phosphate buffer 0.2 *M*, pH 8.0: make up 0.2 *M* stock solutions of $Na_2HPO_4 \cdot 7H_2O$ (53.65 g/ℓ) or $Na_2HPO_4 \cdot 12\ H_2O$ (71.7 g/ℓ), and $NaH_2PO_4$ (27.8 g/ℓ). Mix the solutions in the ratio of 94.7 mℓ of $Na_2HPO_4$ solution to 5.3 mℓ $NaH_2PO_4$ solution and use a pH meter to adjust the pH to 8.0 if necessary, by adding more of the $Na_2HPO_4$ to bring the pH up or more of the $NaH_2PO_4$ to bring it down.
3. 0.01 *M* Phosphate buffer, pH 8.0, prepared by diluting the 0.2 *M* buffer and adjusting the pH.
4. High salt buffer: 0.5 *M* NaCl in 0.01 *M* phosphate buffer
5. Gradient maker consisting of two 100-mℓ chambers; the mixing chamber equipped with a stirrer; gradient makers are available (LKB, Pharmacia) or may be simply constructed using two cylindrical vessels or beakers of the same shape and size, connected by a piece of tubing. The first (low ionic strength) buffer is in the vessel which is connected to the column, and this vessel is stirred. As buffer flows from this chamber to the column, hydrostatic pressure causes the high ionic strength buffer to flow through the connecting tube. The ionic strength, thus, increases gradually during the column run.
6. Protein to be fractionated: this will be either ascitic fluid, dialyzed against the 0.01 *M* phosphate buffer (two changes of 500 mℓ each, overnight) or, preferably, PEG precipitate (see previous section) redissolved in the 0.01 *M* phosphate buffer.

*2. Procedure*

Pack a column with a bed volume equal to the starting volume of ascitic fluid (see note 1). We generally work with batches of 5 to 20 mℓ of ascitic fluid. Pack the column with a slurry of cellulose, according to the manufacturer's instructions. Wash the column with 0.2 *M* phosphate buffer until the pH of the eluate is 8.0 (200 to 300 mℓ buffer, at 90 mℓ/hr). Wash with 0.01 *M* buffer until the conductivity of the eluate is the same as the washing buffer (12 mS). If a conductivity meter is not available allow 25 bed volumes of buffer to flow through the column to ensure equilibration. Apply the protein and elute with 0.01 *M* phosphate buffer (one bed volume) followed by a gradient formed from the 0.01 *M* buffer and the high salt buffer. The flow rate should be about 90 mℓ/hr. Collect 5-mℓ fractions and locate the protein fractions by monitoring optical density. Immunoglobulin elutes before other serum proteins and the monoclonal band should be identifiable as a major band. The immunoglobulin may be concentrated using ultrafiltration equipment (Amicon) or by placing the solution in a dialysis bag and putting the dialysis bag into solid PEG 20,000 (0.7 g/10 mℓ of fluid to be removed).

*3. Notes*

1. Scale: the bed volume is approximately equal to the volume of starting ascites. If a preliminary enrichment step has been carried out the bed volume may be reduced, but the difference is small. Thus, if the immunoglobulin is enriched from 10 to 40%, the nonimmunoglobulin protein falls from 90 to 60%, and the column size may be reduced to two thirds of the size that would have been used for the unfractionated ascitic fluid. As discussed in Section III.A, the size of the column relative to the protein load will affect the yield.
2. The preparation will be contaminated with transferrin, unless this protein has been removed by a preliminary PEG precipitation.
3. The column can be run in 1 to 2 hr.

D. Protein A Method[12,13]

*1. Materials*

1. Protein-A-Sepharose CL4B (Pharmacia, Uppsala).
2. Buffer #1: 0.05 *M* tris/0.15 *M* NaCl, pH 8.6. Dissolve 6.06 g tris base and 8.76 g NaCl, in 800 mℓ water. Adjust pH to 8.6 using 10 *N* HCl. Add 10 mℓ 2% sodium azide stock solution and make up to 1 ℓ with water.
3. Buffer #2: 0.05 *M* phosphate/0.15 *M* NaCl, pH 7.0. Dissolve 4.34 g $Na_2HPO_4$, 2.70 g $NaH_2PO_4 \cdot H_2O$, and 8.76 g NaCl, in turn, in 800 mℓ water. Check pH and adjust if necessary to pH 7.0 by adding 1 *N* NaOH or 1 *N* HCl and make volume up to 1 ℓ.
4. Buffer #3: 0.05 *M* citrate/0.15 *M* NaCl, pH 5.5. Dissolve 2.68 g citric acid monohydrate, 10.96 tri-sodium citrate dihydrate, and 8.76 g NaCl in 800 mℓ water. Check pH and adjust if necessary to 5.5 as above. Make up to 1 ℓ.
5. Buffer #4: 0.05 *M* acetate/0.15 *M* NaCl, pH 4.3. Dissolve 6.8 g sodium acetate and 8.76 g NaCl in 800 mℓ water. Add acetic acid (1 *M*) to bring pH to 4.3 and make up to 1 ℓ with water.
6. Buffer #5: 0.05 *M* glycine/0.15 *M* NaCl, pH 2.3. Dissolve 5.6 g glycine-HCl and 8.76 g NaCl in 800 mℓ water. Adjust pH to 2.3 with 5 *M* HCl and make up to 1 ℓ with water.

Note that buffers keep indefinitely if refrigerated, but pH should be checked and if necessary adjusted before each use.

*2. Procedure*

1. Swell 1.5 g protein-A-Sepharose in buffer #1 according to the manufacturers instructions. The bed volume will be approximately 6 mℓ. Pack a small column (a 5-mℓ syringe or an Amicon affinity column will serve this purpose well); wash through with 20 mℓ buffer #1.
2. Adjust pH of antibody solution to 8.6, either by dilution in buffer #1 or by addition of 1 *M* tris base (added with adequate mixing to avoid local pH extremes). The capacity of the column made from 1.5 g protein-A-Sepharose is about 50 mg antibody.
3. Apply antibody to column, and elute unbound protein with buffer #1. A UV monitor may be used to follow elution. Collect 1- to 2-mℓ fractions, and wash through with five bed volumes.
4. Elute with five bed volumes of each of buffers #2 to 5 in succession. The monoclonal antibody should be eluted in one of these buffers, according to its subclass. Collect the appropriate fraction and neutralize immediately if in acid buffer. The eluate may be allowed to drip into a tube containing 1 *M* tris buffer, pH 8.0 (1/10 of the fraction volume).
5. Dialyze or concentrate immunoglobulin as required. Regenerate column by washing in buffer #5 and then reequilibrate in buffer #1. Between uses the column may be stored in buffer #1 in the cold.

Note that the column run will take about 2 to 3 hr.

## IV. ANALYSIS AND QUALITY CONTROL

### A. Protein Concentration

It is often necessary to know the protein concentration in an ascitic fluid or purified preparation. Two simple methods which do not require extensive calibration each time they are used are described in this section. Many alternative methods are available.

*1. UV Absorption*

Proteins absorb light in the UV region of the spectrum, with a peak in the absorption spectrum at 280 nm wavelength and another around 220 nm. The absorption band at 280 nm is due to the aromatic amino acids. Proteins do not all have the same content of aromatic amino acids; consequently, two proteins at the same concentration may absorb to a different degree (i.e., they have different extinction coefficients). However, these differences are relatively minor, and a very approximate determination of protein content can be obtained simply by measuring optical density at 280 nm, against the solvent buffer, and using the "rule of thumb" extinction coefficient of 1.0; i.e., a protein solution at 1 mg/mℓ in a 1-cm pathlength cell will give an optical density at 280 nm (OD/280) of 1.0. This method gives approximate values only, but is rapid and simple and adequate for many purposes. The sensitivity of the method is limited to 0.05 to 0.1 mg/mℓ. If the protein being determined is pure, a more correct value will be obtained by using the extinction coefficient for the protein, rather than the "rule of thumb" value of 1.0. For IgG from most species a value of 1.35 may be used, i.e., a solution at 1 mg/mℓ in a 1-cm pathlength cell will give an optical density of 1.35. Measurement at 220 nm is more sensitive, but many buffers, as well as dissolved oxygen, absorb at this wavelength and the method is, thus, less easy to use.

A chemical method for protein determination (see next section) can be used to measure the protein content of a standard preparation, which allows the construction of a calibration curve of optical density against protein content. This combines the accuracy of the chemical methods with the ease and convenience of the UV absorption method.

*2. Biuret Method*

The biuret method is based on the reaction of the peptide bond with copper ions to form a colored complex which absorbs light at 580 nm.

*a. Materials*

Biuret reagent includes $CuSO_4 \cdot 5H_2O$, 1.8 g; sodium potassium tartrate, 7.2 g; NaOH, 36.0 g. Dissolve each ingredient separately in 200 mℓ water, mix, and make volume up to 1 ℓ. Isotonic saline is NaCl, 8.5 g, in 1 ℓ water. For protein standard 5 mg/mℓ, dissolve 100 mg bovine serum albumin (BSA) in 20 mℓ water.

*b. Method*

1. Make up dilutions of protein standard in 0.4 mℓ:
   - 0.4 mℓ standard + 0 mℓ saline (2 mg protein)
   - 0.3 mℓ standard + 0.1 mℓ saline (1.5 mg protein)
   - 0.2 mℓ standard + 0.2 mℓ saline (1 mg protein)
   - 0.1 mℓ standard + 0.3 mℓ saline (0.5 mg protein)
   - 0 mℓ standard + 0.4 mℓ saline (0 protein)
2. Make up unknowns in 0.4 mℓ, using saline to dilute the sample to 0.5-2 mg total protein in the sample volume.
3. Add 0.4 mℓ Biuret reagent to each tube, mix and allow to stand 30 min at room temperature. Read absorbance at 580 nm, plot standard curve for BSA and read off concentration of unknowns.

*c. Notes*

The Biuret method as described above is suited to protein concentrations from 1 to 5 mg/mℓ and total protein in the range 0.5 to 2 mg. A number of more sensitive variations are available, requiring less sample. A convenient and sensitive variant of the original Biuret assay uses bicinchoninic acid (BCA) to measure the reduced copper ion formed from the reaction of protein with $CuSO_4$. The reagents are available in a convenient, stable form from Pierce Chemical Company, Rockford, Ill. The suppliers recommend protocols for the range 0.2 to 1.2 mg/mℓ (20 to 120 μg total) and 50 to 250 μg/mℓ (5 to 25 μg total protein).

B. Antibody Concentration

*1. Introduction and Rationale*

There are three distinct approaches to the measurement of antibody concentration: titrating the antibody activity, determining the immunoglobulin as an antigen (using antiimmunoglobulin and a serological assay), and measuring the immunoglobulin as a protein. In a polyclonal antibody preparation, determinations of antibody and of immunoglobulin will measure quite different things, but with a monoclonal antibody preparation the amount of monoclonal immunoglobulin is a useful measure of the amount of antibody present. Titration of antibody activity requires the use of a method to detect interaction with antigen (for instance, the screening test) in a quantitative way, and will not be discussed here. It should be pointed out, however, that precision is poor when the quantitative assay is carried out in serial doubling dilutions. Reproducibility is at best ±1 dilution, and this allows a fourfold difference between the highest and lowest concentration of antibody which may give the same titer. When the antibody assay suffers from this lack of precision, it may be preferable to measure immunoglobulin concentration. However, if the antibody activity has in some way been lost without destruction of the immunoglobulin molecule, immunoglobulin concentration will not give a correct indication of antibody content.

If the monoclonal antibody has already been purified (and activity has not been lost), protein determination will give the antibody concentration immediately. If the antibody is part of a mixture (ascitic fluid or tissue culture supernatant), it is necessary either to measure the total protein and to determine what proportion of the total protein consists of the monoclonal immunoglobulin, or to use a serological procedure to determine the immunoglobulin concentration in the presence of the impurities.

Methods are given below for the serological determination of mouse monoclonal antibody concentration by ELISA and for the physicochemical determination of the monoclonal band as a proportion of the total protein. Both methods involve inaccuracies; the electrophoretic analysis of protein involves the assumption that the amount of dye bound is proportional to the protein concentration, which is a valid approximation within concentration limits. The serological method must be standardized against a monoclonal immunoglobulin of the same subclass, since the antiimmunoglobulin may not recognize all subclasses equally. Thus, a pure immunoglobulin of each subclass must be available if the method is to be used rigorously. For many purposes the finer points may be ignored and approximate values used without concern. There are situations where the concentration needs to be known accurately, for example, when measuring affinity constants, and then the finer points of the determination of immunoglobulin concentration must be taken into account. For everyday purposes we have found the electrophoretic method preferable, as illustrated by the monitoring of purification of monoclonal antibody (Section III).

*2. Determination of Mouse Immunoglobulin by ELISA*

*a. Materials*

1. Goat (or other species) antimouse immunoglobulin, available commercially or prepared as described in Chapter 5, Section V.A.
2. ELISA plates.
3. ELISA reagents and buffers: coating buffer, blocking/washing solution, and detection reagents (see Chapter 6, Section II.B.2).

*b. Procedure*

The assay is performed according to the general ELISA procedure, as described in Chapter 6, Section II.B. An example of an ELISA assay illustrating the standard curve obtained is given in Chapter 6, Section II.B.

1. Sensitize plate with antimouse immunoglobulin in ELISA coating buffer. Concentration of antibody will vary from reagent to reagent, but for a gammaglobulin reagent (not affinity purified) a concentration of 1 to 2 μg/mℓ should be adequate. The use of affinity-purified antibody is preferable, since this avoids competition for binding sites on the plastic. The preparation of affinity-purified goat-antimouse immunoglobulin is described in Chapter 5, Section V.B. Plates may be stored at −20°C for at least 6 months.
2. Wash two times with ELISA washing/blocking solution.
3. Add standards and unknowns: mouse immunoglobulin standard is purified mouse immunoglobulin. Five concentrations, from 0.001 to 1 μg/mℓ, should be used. Incubate for 1 hr at 37°C.
4. Wash two times with ELISA washing/blocking solution.
5. Add mouse immunoglobulin detection reagent: goat antimouse immunoglobulin-

alkaline phosphatase conjugate (Sigma catalog A462 or similar) at a dilution 1/1000 or as determined by preliminary titration; incubate for 1 hr at 37°C.

6. Wash two times in ELISA washing/blocking solution.
7. Add substrate: *p*-nitrophenol phosphate 1 mg/mℓ, in 10% diethanolamine buffer, pH 9.8; incubate for 30 min at 37°C.
8. Stop reaction by addition of 25 μℓ 2 *N* NaOH, and read the optical density at 410 nm; draw standard curve and read off unknowns.

*3. Determination of Monoclonal Immunoglobulin by Zone Electrophoresis and Measurement of Total Protein*

Electrophoresis of ascitic fluid, serum, or concentrated tissue culture supernatant will give a pattern as shown in Figure 2. The monoclonal band is usually clearly defined. By scanning the stained electrophoretic separation the monoclonal band may be expressed as a proportion of the total protein. A separate determination of the total protein (Section A above) allows the calculation of the concentration of monoclonal antibody.

Technical details are summarized below, but the method will depend on the matrix and equipment available and is fully described by suppliers of equipment for electrophoresis. The separation matrix is usually cellulose acetate or agarose, but polyacrylamide gel can also be used. Stains have been developed which approximate to the ideal of density being proportional to the amount of protein, irrespective of the identity of the protein. Electrophoresis scanners are available commercially. Because of the equipment required, the method is particularly attractive if the laboratory already uses it for other purposes. Serum electrophoresis is widely used in diagnostic immunology laboratories (for the detection of monoclonal bands in the serum of patients with myeloma), and most biochemistry laboratories carry out zone electrophoresis in one form or another. If the technique is not available in the laboratory serious consideration should be given to setting it up. Any hybridoma project which will involve purification of the monoclonal antibody will need a method of quantitation, and the electrophoretic method is suitable. However, as pointed out previously, the antibody may be titrated either on the basis of its antigen-binding activity or as a mouse immunoglobulin.

In the author's laboratory electrophoresis of serum or ascites is carried out using Universal Electrophoresis Film (Corning) in a flat-bed electrophoresis apparatus (Behringwerke). The electrophoresis is carried out in 0.075 *M* barbitone buffer, pH 8.6 (20.7 g barbitone; 4.0 g calcium lactate; 131.4 g sodium barbital; 10 g sodium azide in 10 ℓ — dissolve barbitone in 1 ℓ with warming and stirring and add this solution to the other ingredients dissolved in 5 ℓ of water; make up to 10 ℓ; check pH and adjust if necessary). One to 2 μℓ protein samples are applied and electrophoresis is carried out for 45 min at 200 V (Shandon Vokam or similar powerpack). The plates are removed from the apparatus and fixed in 1.25% picric acid in 17% acetic acid for 5 min. (WARNING: picric acid, if dry, is explosive.) The plates are washed in alcohol, blotted under pressure for 10 min, and air dried in front of a hot air blower. The plates are then stained with 0.5% Coomassie blue (Brilliant blue R, Sigma) in 10% acetic acid/45% methanol for 1 min and destained in 10% acetic acid/45% methanol. The plates are finally rinsed in water and dried again in front of a hot air blower.

## C. Monoclonality

*1. Introduction and Rationale*

The various applications of monoclonal antibodies rely on monoclonality, and the interpretation of results would be greatly complicated if a preparation turned out to contain a mixture of antibodies. The importance of verifying monoclonality hardly needs to be emphasized; but verification is not straightforward.

What do we mean, in this context, by monoclonality? There should be only one antibody secreted. Furthermore, all the cells of the population should secrete that antibody, since a nonsecreting subpopulation may increase in proportion during serial passage, leading to loss of antibody titer. It is not sufficient to define a monoclonal antibody by saying that the cells producing the antibody must have arisen from a single cell, because subsequent mutations compromise the monoclonality, in the terms defined above. In particular, mutations which give rise to cells which either cannot make or cannot secrete antibody are relatively common. One specific issue is the secretion, by a monoclonal hybridoma, of mixed immunoglobulin molecules consisting of light and heavy chains from both myeloma and lymphocyte fusion partners. Although the cell population is monoclonal, the antibody molecules are not all the same, and this is a source of difficulty. As indicated in Chapter 3, now that myeloma lines which have lost the ability to code for their own immunoglobulin genes are available, there is no reason to continue to use the earlier lines, which lead to hybridomas secreting mixed molecules.

Monoclonality is defined, for our purposes, partly in terms of the antibody and partly in terms of the cell population. How do we verify monoclonality? There is no method for proving monoclonality; there are several different tests of polyclonality. The usual approach is, therefore, to carry out some of these procedures and assume that the antibody is monoclonal in the absence of evidence for polyclonality.

### 2. Tests on the Antibody

#### a. Chain Composition

Section D below describes the identification of the light and heavy chains in a monoclonal antibody preparation. If, using satisfactory reagents and procedures, more than one heavy or light chain are detected, the antibody is not monoclonal. It should be emphasized that this identification must be done on tissue culture material; ascitic fluid and serum will contain all the immunoglobulins synthesized by cells from the host mouse. The control mechanisms of immunoglobulin gene rearrangement ensure that a single cell cannot produce two light chains; however, since most hybridomas produce kappa light chains, the finding of only one light chain is not a very discriminating test for monoclonality. The mechanisms of heavy chain gene control and RNA splicing do allow a single cell to produce multiple heavy chains. However, hybridomas secreting multiple heavy chains are rare if they occur at all, and the detection of more than one heavy chain must be seen as a strong suggestion that the culture consists of a mixture of different hybrids.

#### b. Electrophoretic Identification of Chains

Isoelectric focusing: principles — Immunoglobulin molecules differ in isoelectric point, reflecting, in part, sequence differences in the variable regions. Monoclonal immunoglobulin may be expected to be relatively homogeneous. Isoelectric focusing may be carried out on the intact immunoglobulin or on the separated light and heavy chains after reduction of intrachain disulfide bonds. The method has been used to detect the different chains produced by a hybrid of an immunoglobulin-producing myeloma and a spleen cell. The value of the method in establishing monoclonality is, however, limited by the fact that a monoclonal immunoglobulin can give multiple bands, due to posttranslational changes such as deamidation and glycosylation. This heterogeneity is slight compared with that seen with polyclonal preparations but, nevertheless, makes interpretation complex. A further complication is that hybridoma cell lines may secrete proteins in addition to their principal protein product, immunoglobulin; thus, nonimmunoglobulin bands may be seen.

Isoelectric focusing: method — Isoelectric focusing is carried out on radiolabeled antibody prepared by intrinsic labeling in vitro. Hybridoma cells are cultured at $5.10^5$ cells/mℓ for 16 hr in leucine-free RPMI medium supplemented with dialyzed FCS, HT, antibiotics, and 5 μCi $^{14}$C-leucine per milliliter. The culture supernatant is concentrated 20-fold and dialyzed against ampholine for isoelectric focusing. The sample is applied to the anodal end of a pH 5 to 9 focusing gel (for a practical discussion of isoelectric focusing see "Further Reading" at the end of the chapter). If the light and heavy chains are to be focused, rather than the whole immunoglobulin, the following method may be used (Hoffman et al.[14]). The concentrated culture supernatant is subjected to gentle reduction and alkylation. The concentrated supernatant is dialyzed against 0.66 *M* Tris/HCl buffer, pH 7.7, and dithiothreitol (0.15 *M* in the same buffer) is added (for example, 0.1 mℓ dithiothreitol solution to 0.6 mℓ concentrated culture supernatant). The reaction should be carried out in the absence of oxygen, by displacing the air by bubbling nitrogen through the solution. After incubation for 1 hr at room temperature the free sulfhydryl groups produced are alkylated by adding 0.3 *M* iodoacetamide in the same buffer (0.1 mℓ for the example given). The pH is maintained between 7 and 8 for 1 hr at room temperature. The light and heavy chains are likely to reassociate noncovalently. To prevent this, propionic acid is added (56 μℓ of 1 *M* acid in the example given) and the mixture is dialyzed against two changes of 10 *M* urea. The dialyzed material is applied to a pH 5 to 9 focusing gel containing 8 *M* urea.

Electrophoresis — Electrophoretic analysis lacks the resolving power of isoelectric focusing, but is, nevertheless, useful in testing for polyclonality. Electrophoretic analysis, without reduction, as carried out to examine ascitic fluid prior to purification (Sections III and IV.B.3 above) can provide useful information on monoclonality. The pattern is expected to reveal a dominant band in the gamma region. Two major bands would be interpreted as suggesting mixed clones; lack of a dominant band simply indicates that the immunoglobulin content is low.

### *3. Analysis of the Hybridoma Line*

The hybridoma cells contain initially two complete sets of chromosomes; early during culture cells lose chromosomes and eventually stabilize with a chromosome number intermediate between diploid and tetraploid. Since this process is essentially random the cell population will not be uniform in DNA content. If the hybridoma is cloned after the chromosome number has stabilized, a population uniform in DNA content should result. Flow cytometric analysis of the cloned hybridoma should, thus, reveal a monoclonal pattern (Figure 5A), with a sharp G/0-G/1 peak and a smaller G/2-M peak with exactly twice as much DNA as the G/0-G/1 peak, and a flat S-phase region between the two peaks. If there are two or more populations with different DNA content, there may be multiple peaks, shoulders, or broadening of peaks (Figure 5B). The technique is simple (if a flow cytometer is available), but limited in that two populations with very similar DNA content will not be distinguished, and very small subpopulations may not be detected. The technique has proven useful, in the particular context of human-human fusions, in distinguishing between hybridomas (which should be hyperdiploid) and EBV-transformants (which should be diploid).[15]

### *4. Effect of Recloning*

If there is any reason to doubt the monoclonality of an antibody or of the hybridoma line secreting the antibody, it should be recloned. The cloning experiment can yield useful information on the monoclonality of the preparation before cloning. Thus, if

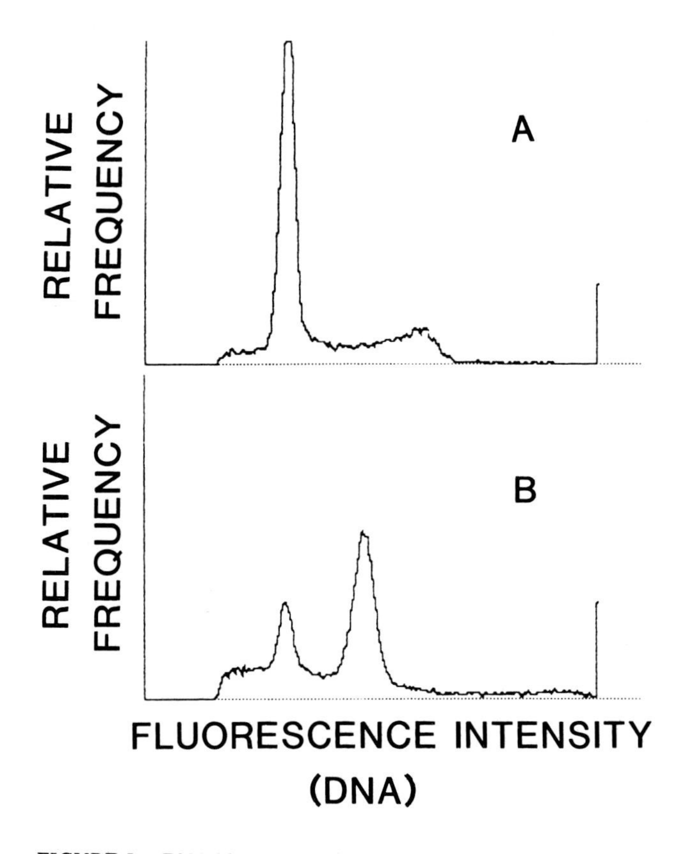


FIGURE 5. DNA histograms of cloned (A) and uncloned (B) hybridoma cell populations. The uncloned hybridoma contains at least two different populations, while the cloned hybridoma shows a DNA distribution characteristic of a monoclonal population — a G0/G1 peak, a smaller G2/M peak with twice as much DNA as the G0/G1, and the intervening S region.

the clones all produce antibody of the same specificity, this result not only yields cloned material, but also some reassurance about the monoclonality of the initial line. If differing specificities are obtained, or if a proportion of clones turn out to be negative, then quite clearly cloning was necessary. One specific situation in which recloning is worth carrying out to confirm monoclonality is when an antibody preparation has a surprising spectrum of reactivity, such that it reacts with two antigens which would not be thought likely to cross react. This is discussed further in the following example.

*5. Example: Monoclonality of FMC3*

FMC3 is a monoclonal antibody produced relatively early in studies of leukocyte antigens. It was found[16] to react with a proportion of human T lymphocytes and a proportion of B lymphocytes. Such a reactivity, with only some members of two functionally distinct cell populations, was regarded as surprising (though more examples of this phenomenon are now known). An alternative explanation was that the preparation contained two antibodies, one against a T-cell subpopulation and the other against a B-cell subpopulation. To verify the specificity of the antibody and, hence, the interesting distribution of the antigen, rather more was required than the usual statement that the hybridoma had been cloned. The following evidence was obtained:

1. Only IgG3 and kappa chains could be identified in the antibody preparation.
2. Isoelectric focusing of radiolabeled FMC3 (prepared by growing the cells in $^{14}C$-labeled amino acids) revealed only one heavy chain. Light chains were not analyzed, because the fusion partner was a myeloma capable of making its own light chain; thus, the hybridoma would be expected to make two light chains.
3. FMC3 was subcloned and two clones tested extensively. The antibodies had the same specificity as each other and as the material produced before recloning. Furthermore, continuous serial passage of the hybridoma over 4 months did not lead to any change in the specificity of the antibody secreted. If the culture was, indeed, a mixture, it would be surprising if the composition of this mixture remained unchanged through approximately 80 passages.
4. The hybridoma line gave a simple DNA pattern, with no evidence of a mixed population.

As a result of these experiments it was concluded that FMC3 was, indeed, monoclonal.

### D. Immunoglobulin Chain Identification

#### *1. Introduction*

Identification of the heavy and light chains secreted by a particular hybridoma is achieved by immunological screening for epitopes characteristic of the different heavy chains (in the mouse: mu, epsilon, alpha, delta, gamma 1, gamma 2a, gamma 2b, and gamma 3) and the two light chains (kappa and lamda). There are several different immunological techniques which may be used for this determination, but they all depend equally on having a set of antisera specific for the different heavy and light chains. These are available commercially, and the quality of these reagents has improved markedly in recent years as the demand for mouse immunoglobulin typing reagents has increased.

The procedure described here was published by Gardner[17] and we have adopted it for routine use unchanged. However, with some knowledge of ELISA methodology (see Chapter 6, Section II.B) it may readily be adapted to meet special circumstances (see notes).

#### *2. Method*

##### *a. Materials*

These include ELISA equipment and buffers as described in Chapter 6, Section II.B. Round-bottomed PVC ELISA trays may be used, since the color is not determined quantitatively. Antibody against mouse immunoglobulin (first antibody, capture antibody) is polyspecific (i.e., will bind all classes and subclasses of mouse Ig); although affinity-purified antimouse immunoglobulin is preferable, a purified immunoglobulin fraction from a high-titer serum may be used. The preparation of goat-antimouse immunoglobulin and its affinity purification are described in Chapter 5, Section V. The concentration needed will depend on the titer of the serum, but a level of around 20 $\mu g/m\ell$ should give a result.

More materials include a set of heavy- and light-chain-specific sera (the typing sera) — we use rabbit antisera supplied by Miles Laboratories. The set contains eight typing sera (antikappa, lamda, the four gamma subclasses, alpha, and mu) and a normal rabbit serum control; enzyme-conjugated antibody against immunoglobulin of the species used for the typing sera — in this case rabbit. We use urease-conjugated sheep-antirabbit antibody together with urease substrate solution (CSL, Melbourne).[18] The antibody is diluted to 1/80 in wash buffer containing 0.25% FCS, 0.25% normal goat serum if needed (see notes), and 0.02 *M* sodium azide. The alkaline phosphatase detection system described in Chapter 6, Section II.B, may also be used, although the urease system is particularly suited to nonquantitative applications.

*b. Procedure*

1. Add 100 μℓ of the capture antibody (polyspecific antimouse immunoglobulin) per well of the microtiter tray, seal tray with Parafilm, and incubate at 37°C for 2 hr. Recover unbound antibody (which may be reused at least twice).
2. Wash three times in wash buffer and store plates, with the buffer still in the plate, at −20°C until required.
3. Thaw out stored plate (or select freshly prepared plate) and wash further two times in wash buffer.
4. Add 50 μℓ of test or control hybridoma supernatants; nine wells per antibody. Seal plate and incubate at 37°C for 30 min.
5. Wash five times in wash buffer.
6. Add 50 μℓ of prediluted typing serum to each well. Seal plate and incubate 30 min at 37°C.
7. Wash three times in wash buffer.
8. Add 50 μℓ conjugated antiglobulin, prediluted.
9. Incubate 30 min at 37°C.
10. Wash three times in wash buffer.
11. Add 100 μℓ urease substrate solution, reseal, and incubate 30 min at 37°C.
12. Read plate — positive wells show transition from yellow to purple.

*3. Notes*

1. If the typing antibody kit does not contain instructions for dilution, the optimal dilution must be determined by preliminary experiments with known hybridomas to ensure that, at the concentrations used, the typing sera are specific.
2. The conjugated antibody may cross react with the first antibody, in this case a goat immunoglobulin. This cross reaction may be prevented by including goat immunoglobulin in the buffer used to dilute the conjugate, or by ensuring that the conjugate and first antibody belong to the same animal species. If the capture antibody binds either the typing antibody or the conjugate, false-positive reactions will result. Reagent batches must be selected to avoid this cross reactivity.
3. If the typing sera are not entirely specific it may be necessary to titrate each reaction quantitatively. In that situation, alkaline phosphatase would be a better enzyme to use, because it gives a quantitative result, unlike urease.

## V. STORAGE AND STABILITY, DISTRIBUTION

### A. Storage and Stability

The general principles for storage of proteins apply to monoclonal antibodies. They are best stored frozen in aliquots such that samples are not thawed and refrozen any more than necessary. Storage at −80°C may be advantageous, but is relatively expensive. Samples stored at −20°C for several years have shown no obvious deterioration. On the other hand, storage at 4°C for only a few days results in some reduction of activity, and freezing and thawing leads to measurable loss in activity. Different antibody preparations behave differently on storage or freezing and thawing, and it is probable that several factors are involved. The antibody class may be important; there is some evidence that IgM is less stable on storage than IgG. The presence of proteolytic enzymes would certainly be deleterious at temperatures above 4°C, and antibody preparations should never be left on the bench for longer than necessary. Ascitic fluid and sera contain proteolytic enzymes, but heat inactivation reduces the risk of proteolysis. As discussed earlier in this chapter (Section II.B.2.e) heat inactivation is not always possible, and ascitic fluids which have not been heat inactivated should be treated particularly carefully with regard to proteolysis. Such antibodies are best purified to remove proteases. The protein concentration is also a significant factor controlling stability; in general, proteins are more stable at high concentration than at low concentration, and even though high dilutions may be required for use the bulk antibody should be stored undiluted. Dilute antibody preparations will lose significant amounts of activity by absorption onto plastic surfaces, and polystyrene storage vessels are particularly bad in this regard. Polypropylene tubes are preferable.

### B. Distribution

Monoclonal antibody companies can arrange rapid distribution of their products. Research workers sending samples to colleagues need to do so with minimum effort and cost. Regulations governing labeling and transport of biological substances, and quarantine regulations of the recipient country, must be taken into account. The International Air Transport Association (IATA) has issued a document "Dangerous Goods Regulations" and most airlines will consider all biological products to fall in this category.

The best way to ship antibodies is freeze dried, but this does involve equipment and expense. Freeze driers are available with adaptors to take multiple ampules, and a satisfactory alternative is the centrifugal vacuum drier. If "professional" freeze-drying equipment, with a capability for vacuum sealing of the ampule, is available, it should be used. The amount of residual moisture in the ampule affects stability and ease of reconstitution and cannot be controlled unless the ampule is sealed either under vacuum or in a controlled atmosphere. If an ordinary laboratory freeze drier is used, ensure that drying goes to completion and seal the ampules as quickly as possible. Freeze-dried material that does not reconstitute properly contains denatured protein, resulting either from inadequate drying or from denaturation during freezing. Some immunoglobulin molecules, in particular, human IgM, are very readily denatured by freezing from solutions of low salt content. In general, the addition of protein to purified immunoglobulin will reduce denaturation; unless there is any real need to purify, it is better to freeze dry raw ascitic fluid. If a freeze-dried antibody fails to reconstitute fully the solution should be centrifuged at 15,000 × G for 15 min to remove large aggregates before use. If binding of aggregates to Fc receptors is a problem, the material may be ultracentrifuged at 100,000 × G for 90 min.

If antibodies are to be shipped without drying they are best shipped frozen, but this involves great expense, since the whole package, including dry ice and insulation, is large and heavy. Shipping frozen is only cost effective if significant amounts of material can be shipped together; as a method of sending individual samples it is very expensive. A cheaper and usually effective method is to post samples in padded envelopes. The sample should be frozen, or at least cold to start with, and should be in a container with a reliable, leak-proof cap, and sealed into a plastic bag as an extra precaution. Unless the antibody is to be used in cell cultures sodium azide may be added as a preservative, at a final concentration of 0.02 *M*.

Cell lines may be shipped in the frozen state (packed in dry ice) or in culture. Experience indicates that delays are almost unavoidable and frequently lead to complete loss of the dry ice and consequent loss of the cells. Cells in culture will usually survive at ambient temperature for a week. They should be cultured for a few days before dispatch, to ensure viability and sterility, diluted just before dispatch to 1 to 3 × $10^5$ such that the culture flask is full to the brim and the top of the flask securely taped. The flask should then be wrapped in absorbent material and put into plastic bags, packaged with adequate insulation to withstand low temperatures and shock in handling and sent with appropriate labels and forms. Such packaging is adequate for most purposes, but will not meet regulations for potentially infectious material. The consignment should be marked "Do Not Refrigerate". The cargo hold of aircraft can reach low temperatures, and it is sometimes possible to negotiate with the airline to ensure that the cells travel in a part of the aircraft that remains warm.

The receiving laboratory should place the cultures in a 37°C incubator overnight, then gently remove most of the medium and feed with fresh medium. If the media used in the sending and receiving laboratory are different, the cells will probably adjust to the new medium, but this should be achieved slowly, and healthy growth in the original medium should be established first. Thus, when sending out cells, send some spare medium.

## REFERENCES


1. Acton, R. T., Barstad, P. A., and Zwerner, R. K., Progation and scaling-up of suspension cultures, in *Methods in Enzymology,* Vol. 58, Jakoby, W. B. and Pastan, I. H., Eds., Academic Press, New York, 1979, 211.
2. Murakami, H., Masui, H., Sato, G. H., Sueoka, N., Chow, T. P., and Kano-Sueoko, T., Growth of hybridoma cells in serum-free medium: ethanolamine is an essential component, *Proc. Natl. Acad. Sci. U.S.A.,* 256, 1158, 1982.
3. McGarrity, G. J., Detection of contamination, in *Methods in Enzymology,* Vol. 58, Jakoby, W. B. and Pastan, I. H., Eds., Academic Press, New York, 1979, 18.
4. Perlman, D., Use of antibiotics in cell culture media, in *Methods in Enzymology,* Vol. 58, Jakoby, W. B. and Pastan, I. H., Eds., Academic Press, New York, 1979, 110.
5. Kilmartin, J. V., Wright, B., and Milstein, C., Rat monoclonal antitubulin antibodies derived by using a new nonsecreting rat cell line, *J. Cell Biol.,* 93, 576, 1982.
6. Zola, H., Mosedale, B., and Thomas, D., The preparation and properties of antisera to subcellular fractions from lymphocytes, *Transplantation,* 9, 259, 1970.
7. Weissman, D., Parker, D. J., Rothstein, T. L., and Marshak-Rothstein, A., Methods for the production of xenogeneic monoclonal antibodies in murine ascites, *J. Immunol.,* 135, 1001, 1985.
8. Brodeur, B. R., Tsang, P., and Larose, Y., Parameters affecting ascites tumour formation in mice and monoclonal antibody production, *J. Immunol. Methods,* 71, 265, 1984.
9. Hoogenraad, N., Helman, T., and Hoogenraad, J., The effect of pre-injection of mice with pristane on ascites tumour formation and monoclonal antibody production, *J. Immunol. Methods,* 61, 317, 1983.
10. Neoh, S. H., Gordon, C., Potter, A., and Zola, H., The purification of mouse monoclonal antibodies from ascitic fluid, *J. Immunol. Meth.,* 91, 231, 1986.
11. Underwood, P. A. and Bean, P. A., The influence of methods of production, purification and storage of monoclonal antibodies upon their observed specificities, *J. Immunol. Methods,* 80, 189, 1985.
12. Ey, P., Prowse, S., and Jenkin, C., Isolation of pure $IgG_1$, $IgG_{2a}$ and $IgG_{2b}$ immunoglobulins from mouse serum using protein A-sepharose, *Immunochemistry,* 15, 429, 1978.
13. Mishell, B. B. and Shiigi, S. M., *Selected Methods in Cellular Immunology,* W. H. Freeman, San Francisco, 1980, 368.
14. Hoffman, D. R., Grossberg, A. L., and Pressman, D., Anti-hapten antibodies of restricted heterogeneity: studies on binding properties and component chains, *J. Immunol.,* 108, 18, 1972.
15. Zola, H., Gardner, I., Hohmann, A., and Bradley, J., Analytical flow cytometry in the study of hybridization and hybridomas, in *Proc. 6th Australian Biotechnology Conf.,* Doelle, H. W., Ed., University of Queensland, Brisbane, Australia, 1984, 405.

16. Zola, H., Beckman, I. G. R., Bradley, J., Brooks, D. A., Kupa, A., McNamara, P. J., Smart, I. J., and Thomas, M., Human lymphocyte markers defined by antibodies derived from somatic cell hybrids. III. A marker defining a subpopulation of lymphocytes which cuts across the normal T-B-null classification, *Immunology,* 40, 143, 1980.
17. Gardner, I. D., An enzyme immunoassay for rapid isotyping of monoclonal antibodies, *Pathology,* 17, 64, 1985.
18. Chandler, H. M., Cox, J. C., Healey, K., MacGregor, A., Premier, R. R., and Hurrell, J. G. R., An investigation of the use of urease-antibody conjugates in enzyme immunoassays, *J. Immunol. Methods,* 53, 187, 1982.

## FURTHER READING

### Cell Culture

Colowick, S. P. and Kaplan, N. O., *Methods in Enzymology,* Vol. 58, Jakoby, W. B. and Pastan, I. H., Eds., Academic Press, New York, 1979.

Adams, R. L. P., *Cell Culture for Biochemists, Laboratory Techniques in Biochemistry & Molecular Biology,* Burdon, R. H. and von Kippenberg, P. H., Eds., Elsevier/North-Holland, Amsterdam, 1980.

Freshney, R. I., *Culture of Animal Cells. A Manual of Basic Technique,* Alan R. Liss, New York, 1983.

### Protein Separation and Analysis

Johnstone, A. and Thorpe, R., *Immunochemistry in Practice,* Blackwell Scientific, Oxford, 1982.

Scopes, R. K., *Protein Purification,* Springer-Verlag, New York, 1982.

Work, T. S. and Work, E., Ed., *Laboratory Techniques in Biochemistry & Molecular Biology,* North-Holland, Amsterdam.

Vol 1: Part I, Gordon, A. H., Ed., Electrophoresis of proteins in polyacrylamide and starch gels.

Part II, Fischer, L., An introduction to gel chromatography

Part III, Clausen, J., Immunochemical Techniques for the identification and estimation of macromolecules.

(Available in one volume or 3 separate parts).

Vol 2: Part II, Peterson, E. A., Ed., Cellulosic Ion exchanges.

Vol 11, Righetti, P. G., Ed., Iso-electric focusing: theory, methodology and applications.

Pharmacia Fine Chemicals produce a range of booklets to go with their separation materials. These are full of practical information, and are available free of charge.

Chapter 5

# USING MONOCLONAL ANTIBODIES: CELL AND TISSUE MARKERS

## I. INTRODUCTION

One of the principal applications of monoclonal antibodies has been in the identification, enumeration, localization, and isolation of individual types of cells from blood or solid tissues. As cells differentiate to carry out their specific functions, different genes are expressed, and the molecular composition of a cell, therefore, reflects its differentiation state. A molecule which is expressed selectively by a particular cell type can serve as a marker for that cell type. Monoclonal antibodies against such a marker molecule can be used to identify, count, find (in tissue), and isolate the cells carrying the marker. The most useful differentiation markers are on the outer cell membrane, since they can be identified without making the cell membrane permeable.

It is likely that a molecule which serves as a differentiation marker is intimately involved in the specialized function of the cell. In many cases we do not know the function of the molecule concerned, and we do not need to know its function in order to use the molecule as a marker.

Although differentiation markers distinguish members of a related family of cells, they may be found in quite different cell lineages. Thus, the CD3 antigen is found in the brain and on T cells, but not on B cells; the CD9 antigen is found on specific parts of the kidney tubular apparatus and on immature, but not mature, B lymphocytes.

Another consideration which should limit our expectations of specificity is that the particular molecular structure detected by a monoclonal antibody (the antigenic epitope) may occur on several unrelated molecules. This would particularly be true of carbohydrate epitopes, which may be found attached to several distinct proteins and to lipids.

In spite of these reservations, monoclonal antibodies have proved extraordinarily useful in the study of cells and tissues, and have revolutionized some of the sciences which depend most heavily on such studies.

The techniques used in applying monoclonal antibodies to cell and tissue studies depend either on a color reaction to detect antibody binding (identification, enumeration, and localization), or on a physical process to separate cells which have bound the antibody from cells which have not. Procedures will be described in this chapter for immunofluorescence (including flow cytometry), immunoenzymatic staining of tissue, and staining with particulate markers. Staining methods suitable for cell suspensions and tissue sections will be described, for use in light or electron microscopy. Methods for the simultaneous detection of two markers will be discussed. Antibody-mediated cytotoxicity will be described, both as an analytical and as a preparative technique. Cell separation procedures based on monoclonal antibody affinity will be described, including flow sorting, panning, and magnetic separation, and the evaluation of the effectiveness of separation will be considered.

## II. IDENTIFICATION AND ENUMERATION OF CELLS IN SUSPENSION

### A. Immunofluorescence

#### *1. Principles*

Fluorescence is a physical process whereby a dye absorbs light at one wavelength and emits light at a higher wavelength. Early fluorescence microscopes used a light source

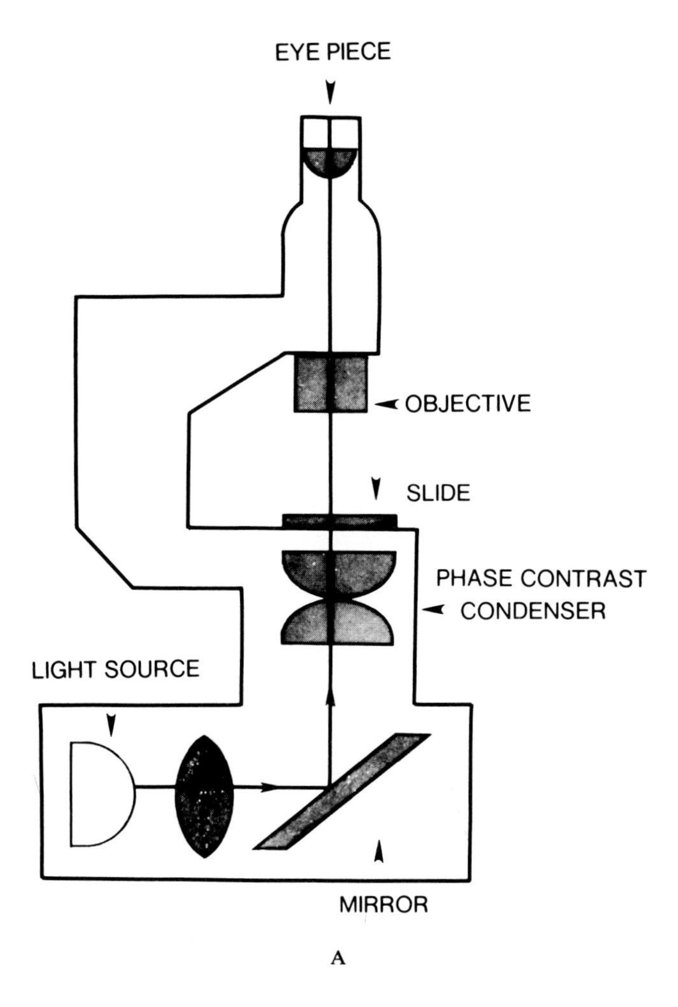


FIGURE 1. Fluorescence microscope showing optical path for phase contrast (A) and fluorescence (B). Note that in fluorescence the incident light beam arrives at the sample from above (epi-illumination), and the emitted light (fluorescence) travels back up the same path, initially. The dichroic mirror reflects light below a certain wavelength, but transmits light of higher wavelengths. This mirror, together with filters, allows fluorescence emission through, but prevents back-scattered light, of the exciting wavelength, from reaching the eyepiece.

which produced invisible UV light. Tissue stained with an appropriate fluorescent dye produced visible blue or green light. The result was that only tissue stained with the dye was visible; the rest was black. Modern fluorescence microscopes achieve the same effect with a source of blue light. Again, fluorescent dyes convert the incident (exciting) blue light to a higher wavelength. Filters are used to block out the excitation wavelength and transmit only the fluorescence emission.

The optical system of the fluorescence microscope is illustrated in Figure 1. Fluorescence optics are best used in conjunction with phase contrast optics, enabling nonfluorescent cells to be seen under white light. Phase contrast optics allow sufficient morphological detail to distinguish, for example, a typical monocyte from a typical lymphocyte, or dead cells from live cells. However, the phase condenser must be constantly checked and adjusted to provide good illumination. Special (expensive) objective lenses are needed to combine fluorescence and phase-contrast optics. Immersion

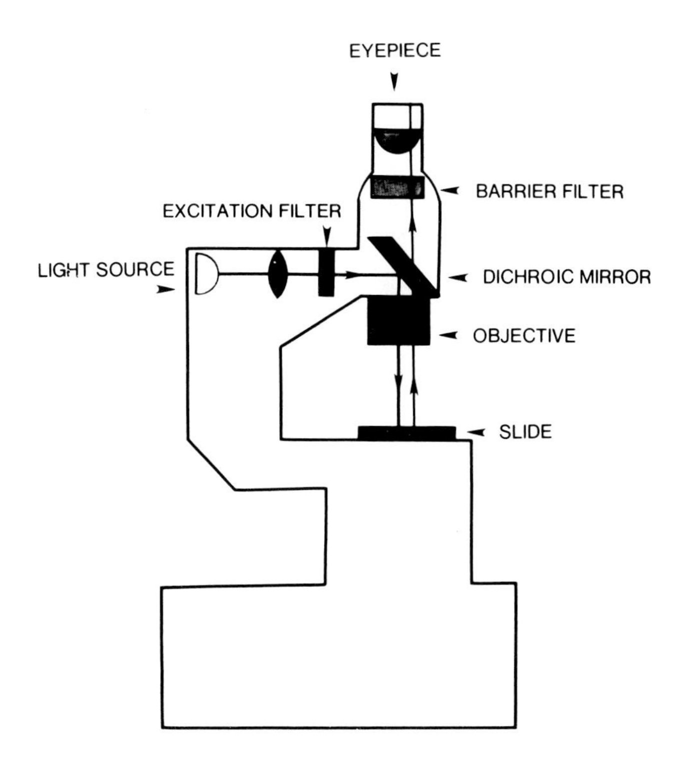


FIGURE 1B

lenses are particularly suited to fluorescence, because they avoid the loss of light at glass/air interfaces which is inevitable with dry objectives. However, the use of immersion fluid adds significantly to the time taken to read a large number of samples. Within a class (dry or immersion) of objectives, lower magnification lenses give higher light intensities, and the same is true for eyepieces. For studies on cells in suspension we use a Leitz Ortholux microscope with a tungsten light source and phase contrast condenser for white light, and a 200-W high-pressure mercury lamp for fluorescence excitation. We use phase/fluorescence (Phaco/Fluotar) objectives, generally a ×40 dry objective and a ×63 dry or ×100 oil immersion lens for high magnification. We use Periplan ×10 or ×6.3 eyepieces and a set of filter blocks as follows: H2 — broad band, allows fluorescein and rhodamine to be read simultaneously, used for single dyes when there is no autofluorescence; L2.1 — narrow band for fluorescein, used when fluorescein emission needs to be distinguished from other dyes or from autofluorescence; N2.1 — narrow band for rhodamine.

Many organic substances fluoresce, but only a few have been exploited as dyes for immunofluorescence. Most work is carried out with fluorescein, which emits green light, and rhodamine, which emits red light. Chemical derivatives of these two dyes are available for conjugation to antibody, and fluorescence microscopes are equipped with light sources and filters designed to use these two dyes at maximum efficiency. Newer dyes have been developed for two-color analysis in flow systems and will be discussed in Section II.G. Techniques for preparing labeled antibody are described in Section V.

There are two distinct methods of using fluorescence to detect antibody binding. Direct immunofluorescence utilizes the antibody directly coupled to fluorescent dye. In indirect fluorescence the unlabeled antibody is reacted with the cells, and then this antibody is detected using a fluorescent dye-labeled antiantibody. The antiantibody, usually referred to as the second antibody, is raised in a different species by immunization with immunoglobulin of the species used to produce the monoclonal antibody.

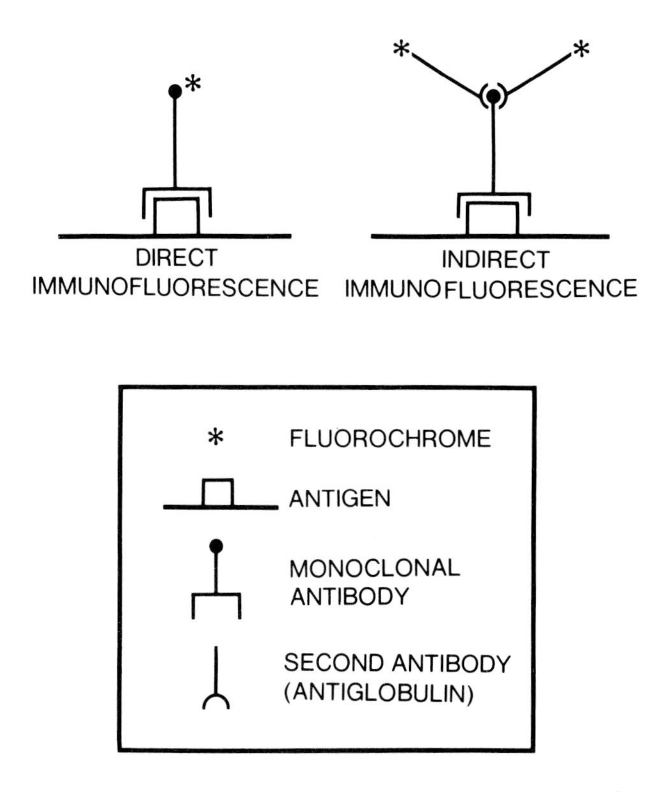


FIGURE 2. Diagrammatic representation of direct and indirect immunofluorescence.

When using mouse monoclonal antibody in indirect fluorescence, the second antibody may be rabbit-antimouse immunoglobulin or goat-antimouse immunoglobulin. The principles of direct and indirect immunofluorescence are illustrated in Figure 2.

Indirect immunofluorescence has a number of advantages over the direct method for most purposes. In particular, a range of mouse monoclonal antibodies can be used with only one labeled second antibody. The indirect technique is also more sensitive, because of the increased number of molecules bound. There are specialized situations, which will be discussed later in this chapter, where direct immunofluorescence is preferable, either in dual-marker analysis or where the second antibody is likely to cross react with antigens on the target cell.

A few alternatives to antiimmunoglobulin reagents have been developed for specialized purposes (Figure 3). Protein A, a product of *Staphylococcus aureus* Cowan 1 strain with a natural affinity for certain immunoglobulins, has been tagged with dye and used for indirect immunofluorescence, replacing the second antibody. The use of protein A with a different detection system, colloidal gold, is described later in this chapter (Sections III.C.3 and III.D). The monoclonal antibody may be derivatized with biotin, and the biotin detected with fluorescent dye-labeled avidin. This approach is useful in dual-markers studies. Alternatively, a second antibody (goat antimouse immunoglobulin) may be labeled with biotin and detected with fluorochrome-conjugated avidin. This system, which is available commercially (Amersham), should give greater sensitivity than the standard indirect immunofluorescence. Biotin-avidin systems are used more extensively with immunoenzymatic detection systems and are discussed in Section III.C.2.

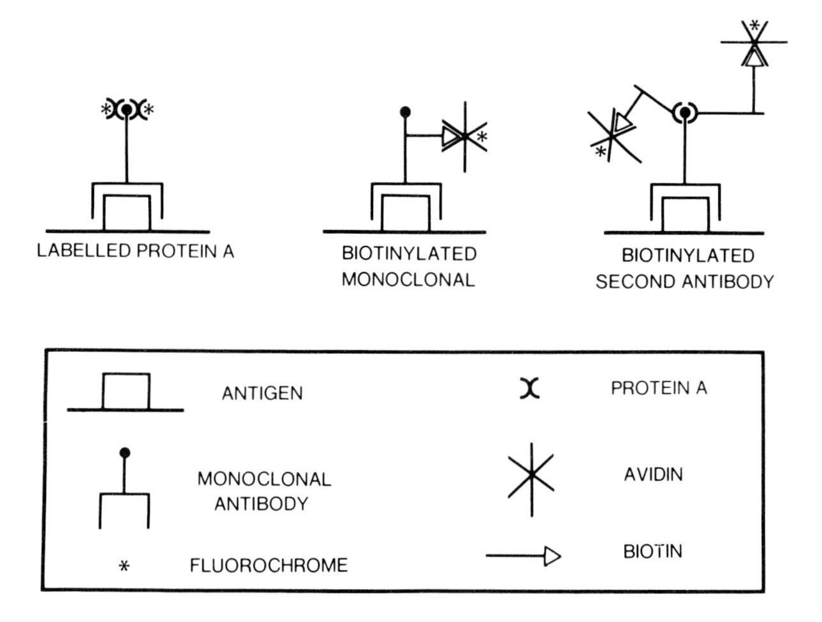


FIGURE 3. Diagrammatic representation of alternatives to labeled second antibody. See text for details.

*2. Method: Indirect Immunofluorescence of Separated Blood Cells with Monoclonal Antibodies*

*a. Warning*

Sodium azide (used in the buffer) is a metabolic poison and forms explosive compounds with heavy metals (e.g., copper). Dispose down the sink with copious amounts of water. Sinks, suction devices, etc. should not have copper or brass fittings. Check with the institute plumber.

*b. Materials*

1. Buffer — Phosphate-buffered saline (Dulbeccos solution A — available in tablet, powder, or liquid form from media suppliers). Alternatively, weigh out NaCl, 10 g; KCl, 0.25 g; $KH_2PO_4$, 0.25 g; and $Na_2HPO_4$, 1.44 g. Make to 0.9 l with distilled water. Adjust pH to 7.2 by adding 1 *M* NaOH or HCl and finally make to 1 ℓ. This buffer is supplemented with sodium azide to a concentration of 0.02 *M* (1.3 g/ℓ) and is referred to as PBS-azide. The azide prevents capping and loss of the antigen (see Note 5). A 100× stock solution of sodium azide (2 *M*; 13 g/100 mℓ) may be used for convenience. Add 10 mℓ/ℓ of buffer.
2. Monoclonal antibody at appropriate dilution — Use culture supernatant undiluted; use ascites at 1/200; use commercially produced monoclonal antibodies as indicated by supplier. See note 4.
3. Second antibody — Fluorescein isothiocyanate (FITC) conjugated to goat-anti-mouse antibody. It is available from several commercial suppliers and may be prepared in the laboratory (Section V.A). It may be necessary to absorb the antibody with insolubilized immunoglobulin of the species from which the target cells are derived, since antiimmunoglobulins tend to have a degree of species cross reactivity. A procedure for absorption is described in Note 6. The antibody is stored frozen in aliquots or in 50% glycerol, which prevents freezing, at −20°C. The antibody should be centrifuged at 10,000 × G for 10 min at 4°C before use, to remove any aggregates, which can bind to Fc receptors on cells, producing false-positive staining.

*c. Procedure*

1. Mix 50 $\mu\ell$ cell suspension at $10^7$ cells per milliliter (in PBS-azide) with 50 $\mu\ell$ monoclonal antibody (or appropriate volume of commercial antibody), containing 0.02 *M* azide.
2. Incubate on ice for 30 min.
3. Dilute with 3 m$\ell$ ice-cold PBS-azide. Centrifuge at 200 × G for 5 min at 4°C. Remove supernatant.
4. Add 20 $\mu\ell$ autologous plasma (or normal human serum) to blood mononuclear cell and buffy coat samples only (see Note 6).
5. Resuspend in 50 $\mu\ell$ FITC — second antibody conjugate.
6. Incubate on ice for 30 min.
7. Wash twice in PBS-azide.
8. Resuspend pellet: for flow cytometry, in 200 $\mu\ell$ PBS-azide; for microscopy, in 20 $\mu\ell$ PBS-azide/glycerol (1 part PBS-azide to 9 parts glycerol).
9. Store samples in the dark at 4°C and read within 24 hr (see Note 2).

*d. Notes*

1. Cell preparation: cells may be prepared for immunofluorescence analysis by a variety of standard methods. Mononuclear cells are prepared from blood by separation on density media (Ficoll-hypaque and similar materials). If the samples are to be analyzed by flow cytometry, whole blood or buffy coat may be used. This allows the examination of mononuclear cell markers without extensive separation, and also permits the study of the other cells of the blood. Blood is collected into anticoagulant tubes and centrifuged at 400 × G for 10 min at room temperature. Most of the plasma is removed (and retained — see note 6). The buffy coat — the cream-colored layer of cells lying over the red-cell layer — is resuspended in a small volume of the supernatant plasma and collected using a Pasteur pipette. The red cells are lysed using Geys hemolytic medium (see Chapter 3, Sections IV.D and VI.C.3. listing 6) or a similar hemolytic medium. The cells are then used in the fluorescence test as above. The mixture contains, apart from lymphoid cells, monocytes, polymorphs, platelets, and some red cells. Analysis of individual cell populations may be done by flow cytometry, using scatter or cell volume parameters to select the individual cell types (Section II.B). The procedure described here is a modification of that described by Hoffman et al.[1] Erythrocytes and platelets are present in such high concentrations in blood that they can generally be studied without removing leukocytes. If enriched populations of red cells are required, the blood can be centrifuged as described above for the collection of buffy coat. The plasma and buffy layers are removed, leaving the erythrocyte-rich pellet. Platelets may be obtained by collecting blood in 3.8% sodium citrate (1 m$\ell$/9 m$\ell$ blood) and centrifuging in polypropylene tubes at 400 × G for 5 min at room temperature. The plasma layer is rich in platelets and may be used directly for immunofluorescence, or the cells may be sedimented before use. Because of their small size, platelets require centrifugation at 800 × G for 5 min to sediment them. Platelets are also fragile, and washing steps should be kept to a minimum. Single-cell suspension may be obtained from tissue by gentle physical dissociation, but enzymatic procedures should be used with caution, since many surface antigens are labile to proteases.
2. Fading due to exposure to light is reduced by storing samples in the cold and dark until they are read. Fading during reading remains a problem, especially if the

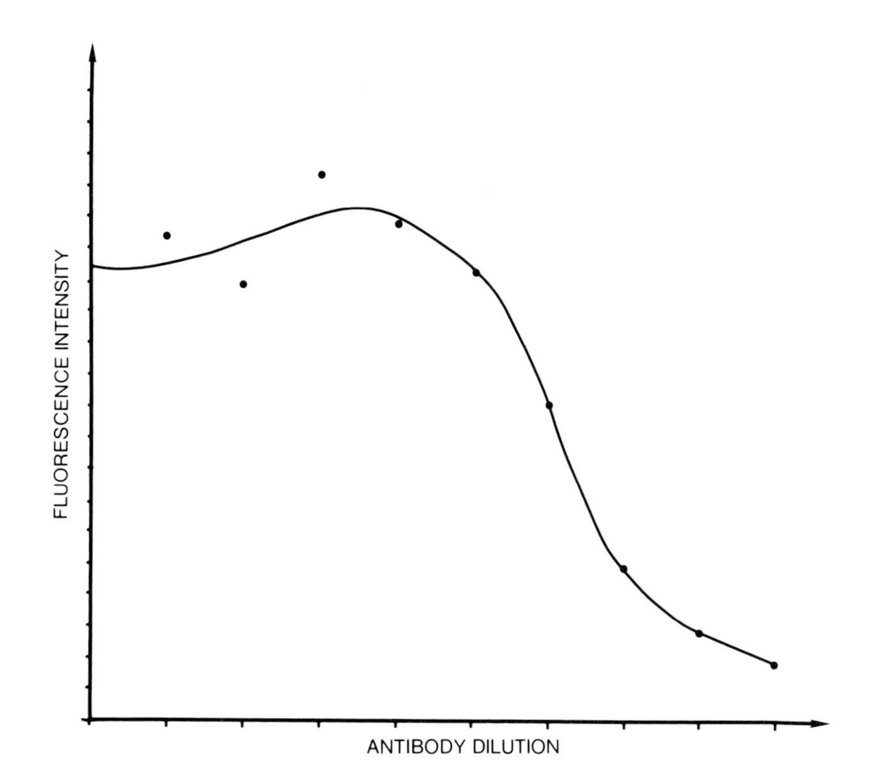


FIGURE 4. Titration curve for a monoclonal antibody. Fluorescence intensity (modal value from a flow cytometric histogram, see Figure 7) is plotted against antibody dilution.

samples are to be photographed. Fading due to light exposure may be reduced by resuspending the cells in PBS: glycerol containing *p*-phenylenediamine[2] at 10 mg/mℓ or *n*-propylgallate[3] at 0.1 to 0.25 *M*.

3. For flow cytometric analysis (see Section II.B), the stained cells may be fixed and stored for up to 1 week before analysis.[4] Procedure: after step (9), resuspend cells in 200 $\mu\ell$ 1% w/v paraformaldehyde in PBS pH 7.2 to 7.4. (NB pH must be checked each time before use.) Cover tubes in foil and leave at 4°C. Read within 7 days.
4. Concentration of antibody: the antibodies must be at a saturating concentration, so that the amount of fluorescence is limited by the antigen density and not by either the monoclonal antibody or the second antibody concentration. The appropriate dilution of antibody is established in preliminary titrations, by testing several dilutions of each monoclonal antibody and the second antibody and reading the fluorescence either by eye or by flow cytometry. When a dilution is reached which shows weaker fluorescence than the previous dilution, the antibody is no longer at saturation (Figure 4). To be safe from fluctuations inherent in the technique, the concentration used routinely should allow at least a twofold dilution without loss of intensity (Figure 4). Highly concentrated antibodies (e.g., ascites fluid) sometimes show a prozone. This term is used when high concentrations of antibody give a negative reaction, while diluted antibody is positive. In the case of immunofluorescence, a prozone could result from the presence of sufficient unbound monoclonal antibody to react with the conjugate in solution, preventing its reaction with cell-bound antibody. In such a case two to three washes will be necessary. Prozones can result from more complex causes, but are uncommon in immunofluorescence (Chapter 3, Section VIII.A.3).

5. Incubation time and temperature: as in most association-dissociation equilibria, low temperature favors association and high temperature favors dissociation. However, equilibrium is reached more rapidly at higher temperature. Thus, a good approach to achieving maximum binding might be an incubation at 37°C followed by dropping the temperature to 4°C. However, another factor which has a bearing on the choice of temperature is the movement of antigen in the plane of the membrane, induced by antibody. This movement is seen variously as patching, capping, endocytosis, and shedding of antigen. The end result may be loss of the antigen (antigen modulation), which clearly must be avoided. These processes are minimized by keeping cells at 4°C throughout the test, and by adding sodium azide. Since temperature cannot be varied to accommodate antibodies of different affinity, incubation time may have to be varied. For most antibodies, 30 min incubation at 4°C is adequate, but low-affinity antibodies may not reach equilibrium under these conditions, and longer incubation times may be needed.
6. Nonspecific staining: negative control antibodies must always be included in the test to check for nonspecific binding. There are two principal causes of nonspecific reactivity in immunofluorescence: cross reactivity of the second antibody with a component of the target cell (often immunoglobulin) and uptake of immunoglobulin by cells with Fc receptors. Cross reactivity is a consequence of the structural similarity of immunoglobulins from different species. An antibody prepared in the rabbit against mouse immunoglobulin will react to some degree with human immunoglobulin. The cross reactivity may or may not be detectable, depending on the batch of antibody and the experimental details of the test. Testing for such cross reactivity by a different assay (such as Ouchterlony double diffusion in gel) is not very helpful; the question is, does the antibody cross react under the conditions of the test? Cross-reacting antibody may be removed (since the antibody will not all be cross reactive) by absorption. For example, when working with mouse monoclonal antibodies against human cell membrane antigens, make an immunoabsorbent consisting of human immunoglobulin immobilized on Sepharose (Chapter 6, Section V). The antiimmunoglobulin (preferably before conjugation with fluorescent dye) is passed down the column and the unbound material collected. The column may be regenerated by standard methods (Chapter 6, Section V). Absorbed antibody may contain antigen-antibody complexes (because antigen leaches slowly off the column). These complexes should be removed by centrifugation (see below). Cross reactivity is at its most troublesome when working within a single species. Mouse monoclonal antibodies against mouse alloantigens, tumor antigens, etc. will always, in indirect immunofluorescence tests, give positive results on B cells or cells with acquired surface immunoglobulin. While many workers have "learned to live with" this problem, it cannot be solved, except by using the monoclonal antibody directly conjugated with dye, or with a derivative such as biotin which allows the use of a different labeled probe. Nonspecific staining due to uptake of immunoglobulin through Fc receptors has three distinct components. The cells may be coated with their own species of immunoglobulin, because they have been in contact with serum. This is, in essence, a component of the cross-reactivity problem already discussed. As will be discussed below, the Fc receptors may be intentionally saturated with autologous immunoglobulin. The second component of the Fc-receptor problem is caused by reaction of the monoclonal antibody with Fc receptors. This is a frequent source of difficulty, as has been illustrated in a number of studies.[5,6] The extent of Fc-mediated uptake varies with the immunoglobulin isotype, so that for critical studies negative control antibodies of the same class and subclass as the

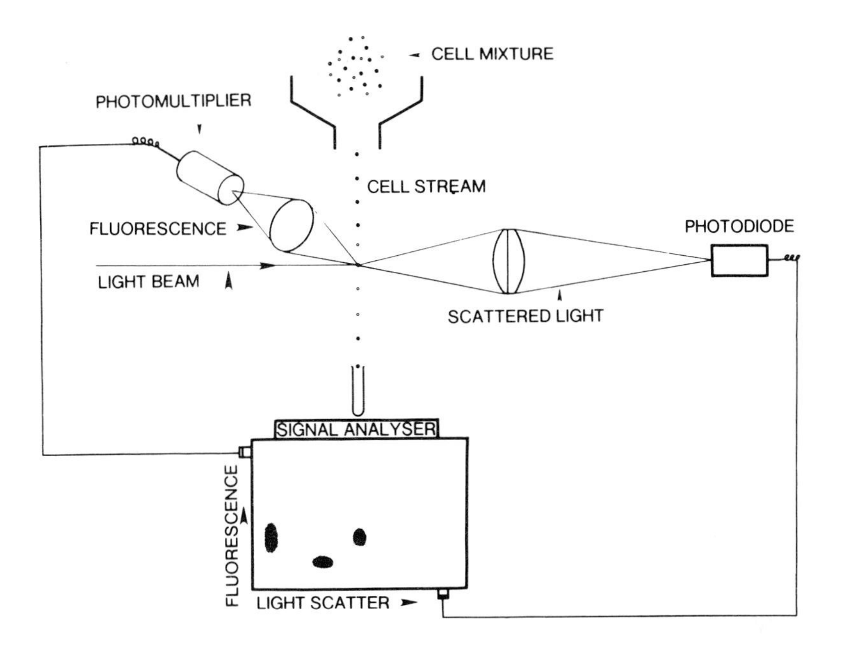


FIGURE 5. Diagrammatic representation of flow cytometry. Cells pass, one at a time, through a light beam. As each cell passes signals (fluorescence, light scatter) are recorded by photo-sensitive devices and amplified, accumulated, and displayed by an electronic signal analyzer.

monoclonal antibody under study must be used. Finally, the fluorescein-conjugated second antibody may be taken up by Fc receptors. For critical work $F(ab')_2$ fragment may be used. Goat immunoglobulin is generally found to give less nonspecific staining of human tissue than rabbit immunoglobulin, presumably, due to a lower affinity for the Fc receptors. Aggregates or complexes react more strongly with Fc receptors than does monomeric immunoglobulin; antibody should, therefore, be centrifuged before use. While spinning at 10,000 × G for 10 min is generally adequate, for critical work ultracentrifugation at 100,000 × G for 60 min is preferable. A partial solution to Fc-mediated uptake is to block Fc receptors with autologous immunoglobulin. This is done by adding autologous plasma, as described in Step 5 of the method. Clearly, this procedure cannot be used where the antigen to be detected is also a serum protein (for example, immunoglobulin), and will not help reduce nonspecific staining if the second antibody cross reacts with immunoglobulin of the target-cell species.

### B. Flow Cytometry: Principles

Flow cytometry is a technique for the automatic and quantitative determination of physical parameters of cells, including fluorescence induced by the attachment of fluorescent dyes. Instead of looking at a number of cells randomly distributed around a microscope slide, the cells are caused to flow in single file through a light beam. The way in which they interact with the light beam provides information about their physical properties and is measured by optical sensors. The general principle is shown in Figure 5.

Operation of flow cytometers is a specialized technique which will not be described in this book. Usually, the instrument is operated by a specialist technician, but the person supplying the samples needs to understand the capabilities of the instrument and the nature of the results in order to make the best use of the technique.

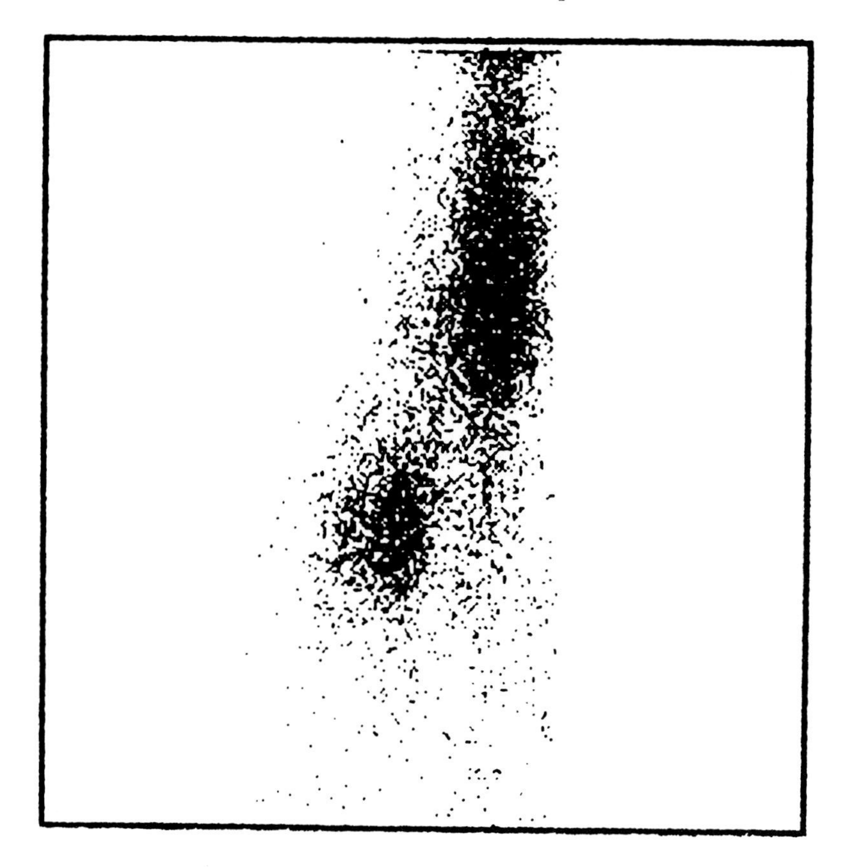

FIGURE 6. Dot-plot display from flow cytometer. Each dot represents a cell, and the coordinates of the dot represent any two selected parameters. In the example shown, light scatter (X axis) and fluorescence (Y axis). In the photograph, there is a large number of cells not reacting with the antibody and a distinct population of cells (light scatter greater) which is positive with the antibody (fluorescence greater).

A flow cytometer collects data on each cell and accumulates the data in a memory. To take a typical example, suppose we pass 25,000 lymphocytes through the flow system. Each lymphocyte will give a scatter signal, detected by a photodiode which is in a direct line with the incident light beam. A beam-stop cuts out the direct, unscattered beam; only light which has been scattered and, thus, deviated slightly from the beam axis will be collected and measured. This pulse of scattered light recorded as each cell passes through the light beam serves to "trigger" the counting system — it informs the instrument that a cell has arrived at the light beam. As each cell traverses the light beam any light given off at right angles to the direction of the light beam is detected by photomultipliers. By using appropriate filters to cut out the wavelength of the incident light, the photomultiplier will record only fluorescence emission. For each cell, therefore, two signals — light scatter and fluorescence — are recorded. The data are recorded in a correlated manner, so that they can be displayed in the form of a dot-plot, in which each dot represents a cell, with the coordinates representing low-angle scatter and fluorescence intensity (Figure 6). Alternatively, the data may be output in the form of independent frequency distributions for each parameter, as illustrated in Figure 7. The frequency distribution is conceptually the same as a histogram, but has the appearance of a smooth curve because the number of data points is large (typically 256). Interpretation of a fluorescence histogram is illustrated in Figure 8.

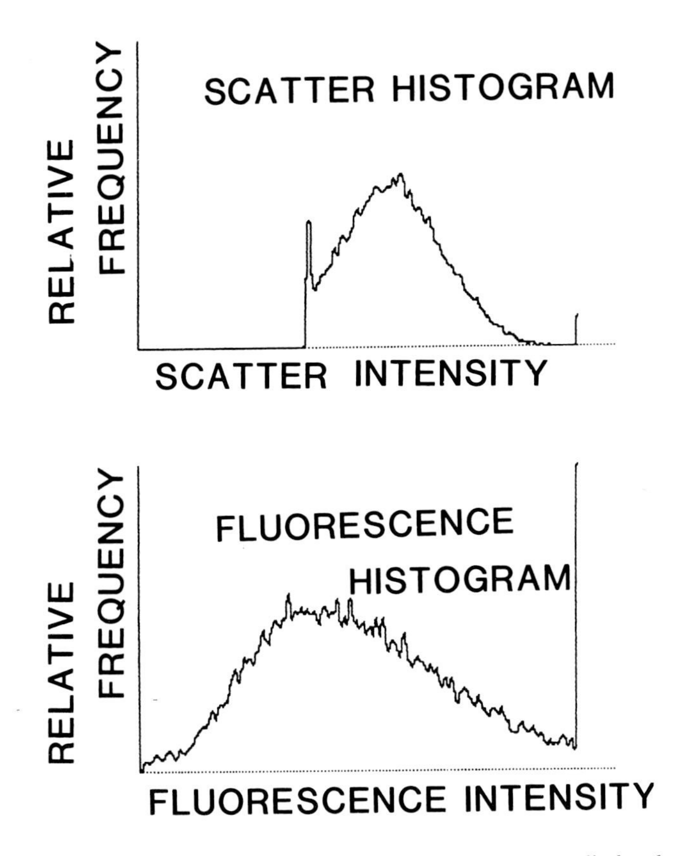


FIGURE 7. Uncorrelated histograms of two parameters, displayed by a flow cytometer.

Apart from enabling the instrument to count each cell as it comes through, the low-angle scatter signal provides some information on the properties of the cell. In approximate terms, the intensity of the low-angle scatter signal is proportional to the size of the cell. Additional properties may be measured, including light scatter at right angles to the incident light beam and, in some instruments, cell volume. The intensity of the right-angle scatter signal produced by a cell depends in an approximate way on its "structuredness". A cell with a rough membrane or a large amount of cytoplasm with many organelles will scatter more than one with a smooth membrane and little structured cytoplasm. Blood neutrophils are readily distinguishable from lymphocytes by right-angle scatter, and a combination of low-angle and right-angle scatter allows red cells, platelets, lymphocytes, monocytes, and granulocytes to be distinguished, albeit with some overlap. Some instruments can use correlated data from four parameters simultaneously, and electronic "gating" enables the user to ask questions such as: "what proportion of cells with low-angle scatter intensity greater than x but less than y, and right-angle scatter intensity between v and w, have fluorescence intensity greater than p but less than q?" This type of analysis is illustrated in Figure 9.

## C. Counting Cell Populations using Monoclonal Antibodies

The method of cell enumeration using flow cytometry is implicit in the general principles discussed above. If a monoclonal antibody stains a proportion of cells in a population, the frequency distribution will be bimodal (Figure 8). The proportion of cells in the brighter peak can be automatically recorded by the instrument. A negative con-

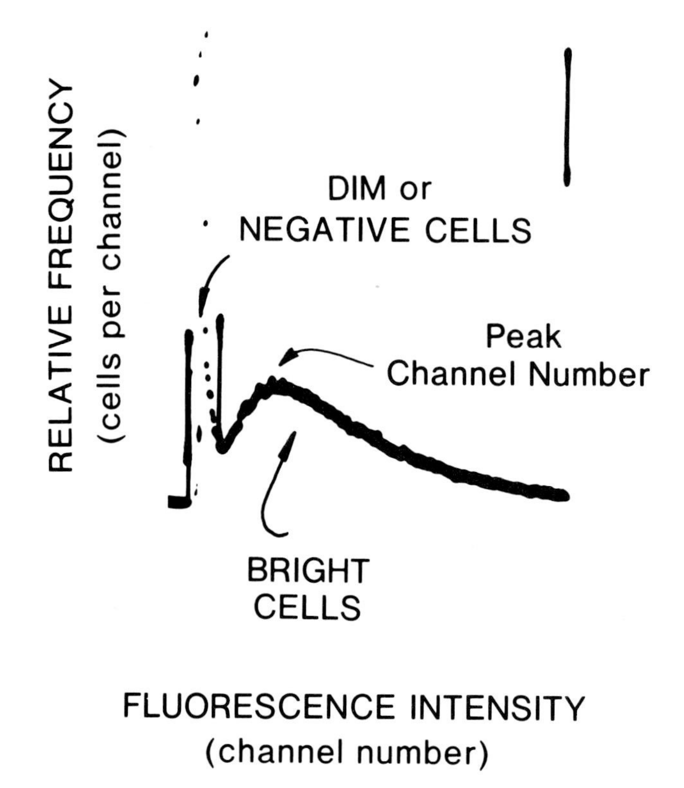


FIGURE 8. Interpretation of a bimodal frequency distribution. The peak position (peak channel number) is also called the modal fluorescence intensity.

trol antibody should always be run in parallel (Figure 10). Flow cytometric analysis has a number of advantages over microscopic counting for the enumeration of the proportion of cells bearing a particular antigen. The precision is much higher in flow cytometry, because it is possible to count a large number of cells rapidly (typically 25,000, as compared with 100 to 200 usually counted in the microscope). Accuracy and objectivity are greater in the flow cytometer, particularly in instances where the fluorescence intensity of the positive cells is not much higher than background fluorescence in the negative controls. If the frequency distribution is bimodal (Figure 8) the decision as to whether a particular cell is positive or not is straightforward. In those instances where the distribution is unimodal and the positive cells form a "tail" the pattern may be easier to interpret as a logarithmic plot (Figure 11). If the logarithmic distribution does not show a clear positive peak an accurate answer is not possible, but nor would it be possible in the microscope to distinguish weak positive from background staining. In such instances, it is possible in the flow cytometer to set the marker at a value that excludes most of the cells in the negative control and defines cells above this selected threshold as positive. This is essentially what is done by eye in the microscope technique, except that it is not possible for the human memory to hold a threshold value constant during the course of reading a set of samples, and the threshold value set by different people varies considerably.

Furthermore, use of the additional parameters available in flow cytometry allows more sophisticated analysis, for example, analysis of lymphocyte markers in the presence of other blood cells. Combining phase-contrast optics with fluorescence does enable the skilled microscopist to carry out the same analysis, but considerable experience

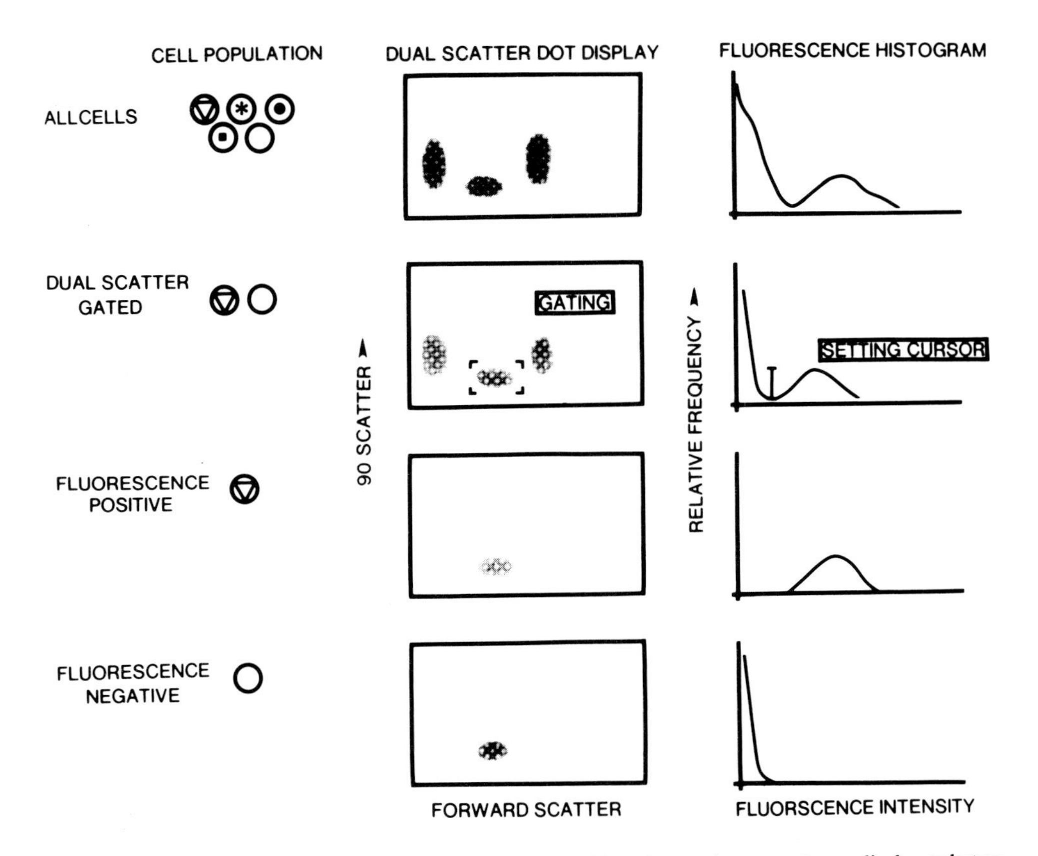


FIGURE 9. Multiparameter analysis by flow cytometry. Although most instruments can display only two parameters at one time, up to four parameters can be stored and used. In the example shown, a complex mixture of red cells, lymphocytes, and granulocytes is analyzed. The top row shows a complex dual-scatter dot display and a fluorescence histogram indicating a proportion of cells positive with the monoclonal antibody under test. In order to analyze the reactivity of the lymphoid cells alone, the portion of the dual-scatter plot known to consist largely of lymphocytes is selected by "gating" (second row from top). Gating instructs the signal analyzer to preclude cells outside the gates from further analysis. The fluorescence histogram of the gated population (second row, right) indicates that a proportion of the gated cells (lymphocytes) react with the antibody. By setting the cursor in the trough between positive and negative cells, the % positive may be determined (automatically, by the instrument). Subsequent analysis, for example, using a second fluorescence channel, or sorting, can be done on the fluorescence-positive or fluorescence-negative cells (third and fourth rows).

is needed to distinguish monocytes from lymphocytes with any reliance on the basis of their appearance under phase contrast. Analysis of complex mixtures is also possible by microscopy, using immunoenzymatic staining and counterstaining for morphology (see Section III).

Flow cytometers are expensive instruments requiring specialized operation, and are not available to all who wish to use monoclonal antibodies to count cell subpopulations. If a flow cytometer is not available, the immunoenzyme methods should be considered (Section III), but immunofluorescence by microscopy is capable of giving useful information. It is essential to use phase-contrast optics to detect nonlabeled cells, since these cannot be seen clearly by normal illumination. Phase-contrast microscopy is only effective if the condenser is properly adjusted, a relatively simple procedure that requires daily attention. Using a good quality microscope properly set up, it is possible to obtain reliable counts of positive and negative cells by immunofluorescence, although the precision and objectivity cannot rival that obtained by flow cytometry. An important consideration is the nature of the information required. Thus,

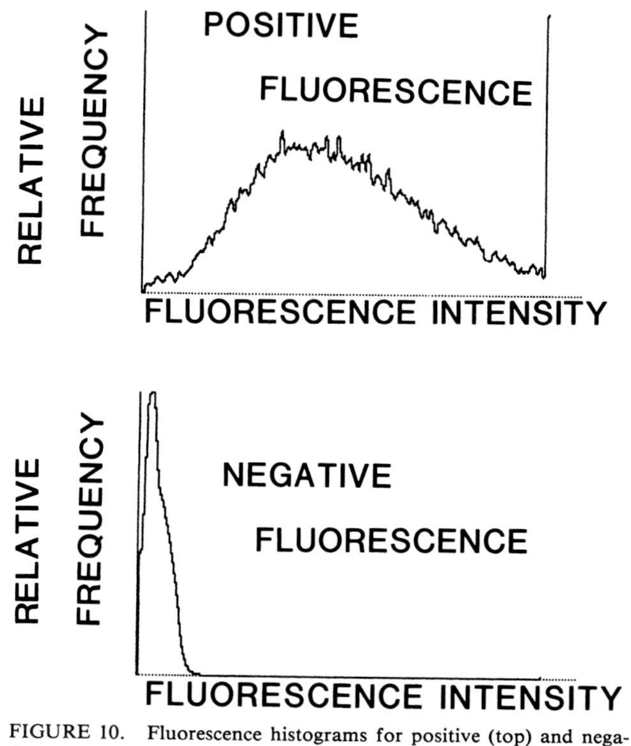


FIGURE 10. Fluorescence histograms for positive (top) and negative (bottom) control antibodies.

when determining the phenotype of a lymphocytic leukemia the question is essentially qualitative: are the cells T cells or B cells, or non-T-non-B? When looking at a remission marrow from the same patient, the number of cells with a particular phenotype is important. Flow cytometry has a significant advantage over microscopy in the second situation, but not in the first.

D. Quantitation of Cellular Antigen

*1. Comparison of Cells within a Population and between Populations*

Within a population of cells which express a particular marker, there is a distribution ranging from cells expressing small amounts to cells expressing large amounts of marker. This distribution is often broad, with the bright cells having five to ten times as much marker as the weakly positive cells. Flow cytometry provides an accurate analysis of the distribution of intensity when the marker is detected by immunofluorescence (Figure 8). In the same way, the expression of a particular antigen by different cell populations may be compared quantitatively. This approach can be used to study variation in antigen expression during the cell cycle, during maturation, or in response to external stimuli.

*2. Antigen Biochemistry*

Quantitative analysis of antigen concentration has been used to determine some aspects of the chemistry of cellular antigens.[7] Cells may be treated with proteases or glycosidases to determine whether a particular antigen is protein in nature, and whether the binding site includes particular sugar residues. If the enzyme treatment is done under carefully controlled conditions the cells remain viable, and reexpression of the antigen can be studied in the presence of inhibitors of protein synthesis or glycosylation, to confirm the nature of the antigen. The rate of turnover of the antigen may be determined similarly.

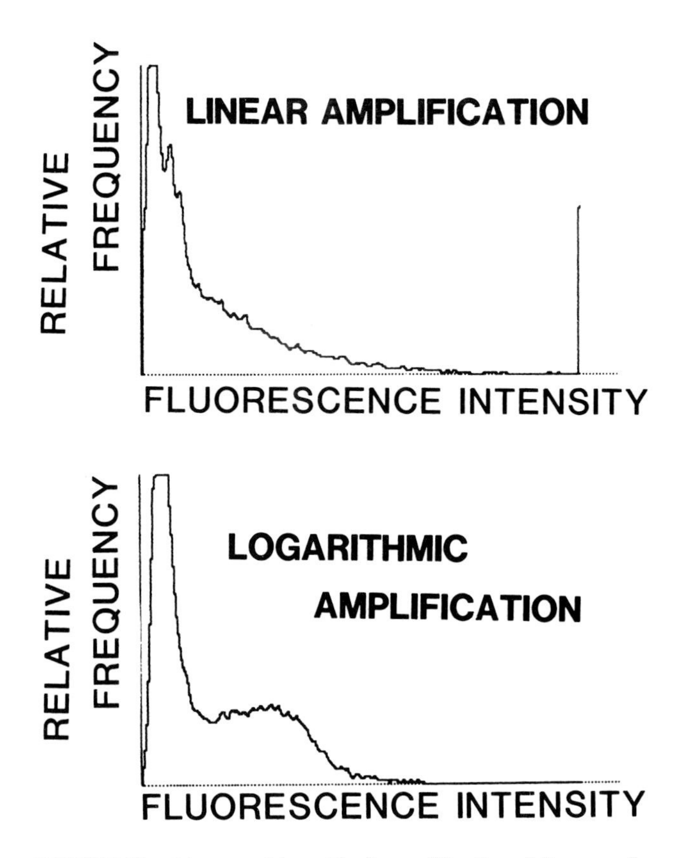


FIGURE 11. Linear and logarithmic amplification of the same fluorescence distribution.

*3. Additive Binding*

When two antibodies react with the same cell a variety of experiments can be carried out to determine whether the antibodies react with the same antigen and with the same or overlapping epitopes. If the two antibodies, each used at saturation, give additive binding they do not compete with each other for the same binding site. If they do compete they are likely to be against the same epitopes or overlapping epitopes. Examples of this type of analysis have been described.[6,7] It is interesting to note that the concept of steric hindrance has changed over the years. It used to be believed that the antibody molecule, being large in relation to the epitope bound, would inhibit binding of other antibodies to different antigens in the vicinity. It is now more common to think in terms of inhibition of binding only to determinants which are on the same molecule and very close to the epitope bound. This can be justified in terms of the conformational flexibility of protein molecules. However, it would be premature to believe that steric hindrance is fully understood.

Deciding whether or not two antibodies react with the same antigen is probably best done by biochemical experiments, such as sequential immunoprecipitation. However, if one of the antibodies modulates the antigen (see Section II.A.2.d. Note 5) it is possible to determine whether the other antibody still binds to the cell, in which case the antigens are independent.

## E. Rosetting Methods

Rosetting methods are based on the use of a particle as an indicator for antibody binding. The method is conceptually similar to immunofluorescence; instead of binding fluorescent dye to the antibody, a particle is used to adsorb the antibody. This antibody-coated particle now binds to antigen-bearing cells. The term rosetting derives from the use of erythrocytes as the indicator particle; white cells surrounded by antibody-coated erythrocytes have the appearance of rosettes. The particles most often used are still erythrocytes, although bacteria,[8,9] dye-labeled latex beads (such as the Quantigen$^{TM}$ T and B cell assay system from Bio-Rad Laboratories), and colloidal gold particles[10,11] have also been used. Colloidal gold is used mainly for electron microscopy (Section III.D), but, with the help of silver intensification, colloidal gold-labeled antibodies, lectins, or protein A provide reagents of high sensitivity (Section III.C.3).

The rosetting methods have a number of potential advantages over immunofluorescence, which make them worthy of consideration. They are claimed to have greater sensitivity than immunofluorescence, although it is always difficult to compare data from different laboratories using different techniques. Rosetting methods are more adaptable to multiple marker studies (Section II.G), to experiments where counterstaining is desirable in order to detect morphological features of the reacting cells, and to preparative use (Section IV.D).

The major difficulty in rosetting methods is achieving effective coupling of the antibody to the particle. Coupling of antibody to erythrocytes has generally been carried out using chromic chloride treatment, a method which appears to have some uncontrolled variables, to the extent that it does not "export" readily from one laboratory to another. Likewise, the methods based on bacteria and latex particles have not achieved widespread use, although their potential advantages would have secured acceptance if they had been easy to set up. The methods which utilize fluorescent latex particles would be particularly useful for two-color fluorescence, but have proved difficult in practice. Nevertheless, rosetting methods have been used effectively for the screening of hybridomas and for studies with monoclonal antibodies, and should be given some consideration, particularly if it is possible to obtain instruction on the procedure from a laboratory using a rosetting method.

### F. Cytotoxicity

#### *1. Principles*

One of the natural functions of antibody is to bind to a foreign cell (bacterial or parasitic) and initiate a reaction cascade in a series of serum proteins known collectively as complement. The reaction, referred to as complement fixation, culminates in lysis of the cell. Complement-mediated cell lysis can be used to detect and quantitate antibody against cell surface antigens, and to remove cells bearing particular markers from a mixture.

To detect cytotoxic antibody, cells are incubated with antibody, and normal serum is added as a source of complement. Lysis of the cell membrane is demonstrated by changes in its permeability — by showing that a dye or radioactive marker which will not normally traverse the membrane does so after lysis. Alternatively, cell lysis may be measured by assessing the ability of cells to divide or metabolize. Four different methods for determining cytotoxicity will be described here, each with advantages in particular situations. Dye exclusion methods depend on the use of dyes which cannot penetrate the cell membrane until it is damaged. The method can be set up in a straightforward way, as it is used for assessing cell viabilities. Used in this way, dye exclusion lacks precision. On the other hand, dye exclusion is used with great precision in tissue-typing laboratories. The procedures used in tissue typing are not simple and require careful control of experimental variables. A simple dye-exclusion method will already be in regular use in most laboratories to check viability of cell cultures. Trypan blue is the dye used most commonly; a number of other dyes may be preferred for a variety of reasons. Fluorescent dyes which bind to DNA are used to detect dead cells in flow cytometry. Ethidium bromide is often used for this purpose because it can be excited by the same wavelength used for excitation of fluorescein, but gives a red emission, which can readily be resolved from the green emission of fluorescein.

A technique which is similar in principle to dye exclusion utilizes fluorescein diacetate. This colorless ester enters cells freely, and is then hydrolyzed to free fluorescein by cytoplasmic esterases. The fluorescein cannot get out of the cell unless the membrane is damaged, and live cells are, thus, stained; dead cells are unstained. This method, if used in conjunction with flow cytometry, is capable of precision and sensitivity when the application requires measurement of the lysis of a minority cell population (as will be described in the example in Section 3 below). Another method which is suited to the analysis of minority target cell populations is the clonal assay, discussed in Section 4 below. This method is restricted to proliferating target cells. Other methods for determining cell proliferation may also be used to determine viability after treatment with antibody and complement. Methods based on the incorporation of radioactive nucleotides or amino acids are capable of great precision.

The most precise method, in terms of antibody titer, is the isotope release method, described in Section 5. This method depends on the use of an isotopic indicator of membrane integrity. The relatively short half-life of the isotope (usually $^{51}Cr$) and the complexity of the technique (bearing in mind the precautions needed in handling the isotope) make isotope release methods attractive only if cytotoxicity is a major interest.

### *2. Dye Exclusion (Trypan Blue)*

The procedure described is one used by the author in the past and now superseded by the fluorescein diacetate method (Section 3). It should be emphasized that there are many variations of dye-exclusion tests, and our procedure may not be optimal. Tissue-typing laboratories would have extensive experience in this type of assay and would use different dyes (particularly, eosin), and Terasaki microplates, which require less of each reagent. The Terasaki plate may be read on an inverted microscope.

#### *a. Materials*

1. Hanks balanced salt solution (HBSS).
2. Ninety-six-well microtiter tray, round-bottomed or, if test is to be read in the tray, flat-bottomed.
3. Target cells, which should have a viability of >90% at the start of the test.
4. Test, negative and positive control sera.
5. Complement, usually rabbit serum.
6. Trypan blue stock solution: 0.2% w/v in water; trypan blue diluent: NaCl 4.25 g/100 mℓ water; trypan blue working solution (made up fresh each day: 4 parts trypan blue stock + 1 part diluent).

#### *b. Method*

1. Add 25 μℓ HBSS to each well.
2. Add 25 μℓ serum sample to first well, dilute across.
3. Add 25 μℓ target cell suspension at $10^7$/mℓ in HBSS.
4. Place in 37°C incubator and leave for 30 min.
5. Add 25 μℓ complement. See Note 5.
6. Incubate further 30 min at 37°C.
7. Remove supernatant by inverting plate and flicking (see Note 3).
8. Add 1 drop trypan blue solution and leave for 2 min.

9. Remove trypan blue by flicking.
10. Add 1 drop HBSS.
11. Either examine under inverted microscope or use a glass rod to resuspend the cells and transfer a drop from each well to a slide for examination under the microscope.

*c. Notes*

1. Dead cells take up the blue dye. The test is designed to give an approximate answer only and the samples should be scanned to find the serum dilution at which approximately 50% of cells are dead. This is the titer of the serum.
2. The test should be read quickly, because there is a gradual increase in the percentage of cells stained. This is not a problem with eosin, which is less toxic to cells than is trypan blue.
3. The procedure for flicking out supernatant requires some experience. The flick should remove most of the liquid without resuspending the cells; if this is not done evenly a second flick may cause loss of cells.
4. Aliquots can be dispensed and sera diluted across using a 25-$\mu\ell$ plunger-type micropipette. Multichannel dispensers, which are designed to dispense into and sample from eight rows of a microtiter plate simultaneously, save a great deal of time, but it is important to check that each tip is on properly. Alternatively, diluters may be used; these devices mix the contents of a well and transfer 25 $\mu\ell$ as they are moved to the next well, and eight can be used simultaneously so as to dilute eight rows in one operation.
5. Complement: all mammalian sera contain the series of enzymes referred to collectively as complement, but not all are equally effective. With rat and mouse antibody, rabbit complement is effective, but bleeds from different animals vary in efficacy and in the nonspecific (background) toxicity. Background toxicity may be reduced by diluting the serum, but efficacy is usually lost beyond dilutions of 1/4 to 1/8. A new batch of complement should be tested to determine the optimum dilution and to check for toxicity (which may vary depending on the target cell used). Complement activity is lost rapidly on storage and is also susceptible to repeated freezing and thawing, so that aliquots suitable for a single experiment should be stored at −80°C. Background toxicity of complement results largely from the presence in the rabbit serum of antibodies which cross-react with the target cell membrane. Although procedures are available for rendering rabbit serum nontoxic, it is generally better to screen a number of rabbit bleeds, or buy a tested batch. Blood from young animals is less likely to be toxic. If the work concerns largely one type of target cell, toxic sera may be absorbed with this cell type on ice for 30 min.
6. Controls: because of nonspecific cell death it is necessary to run a full series of controls, including, particularly, monoclonal antibody without complement (or with heat-inactivated complement) and complement without antibody (preferably with a control antibody, which does not react with the target cell, at the same concentration as the test antibody).

*3. Fluorescein Diacetate Method*

*a. Materials*

1. Stock solution: dissolve 5 mg fluorescein diacetate (FDA) in 1 m$\ell$ acetone. Wrap container in aluminum foil and store at 4°C.

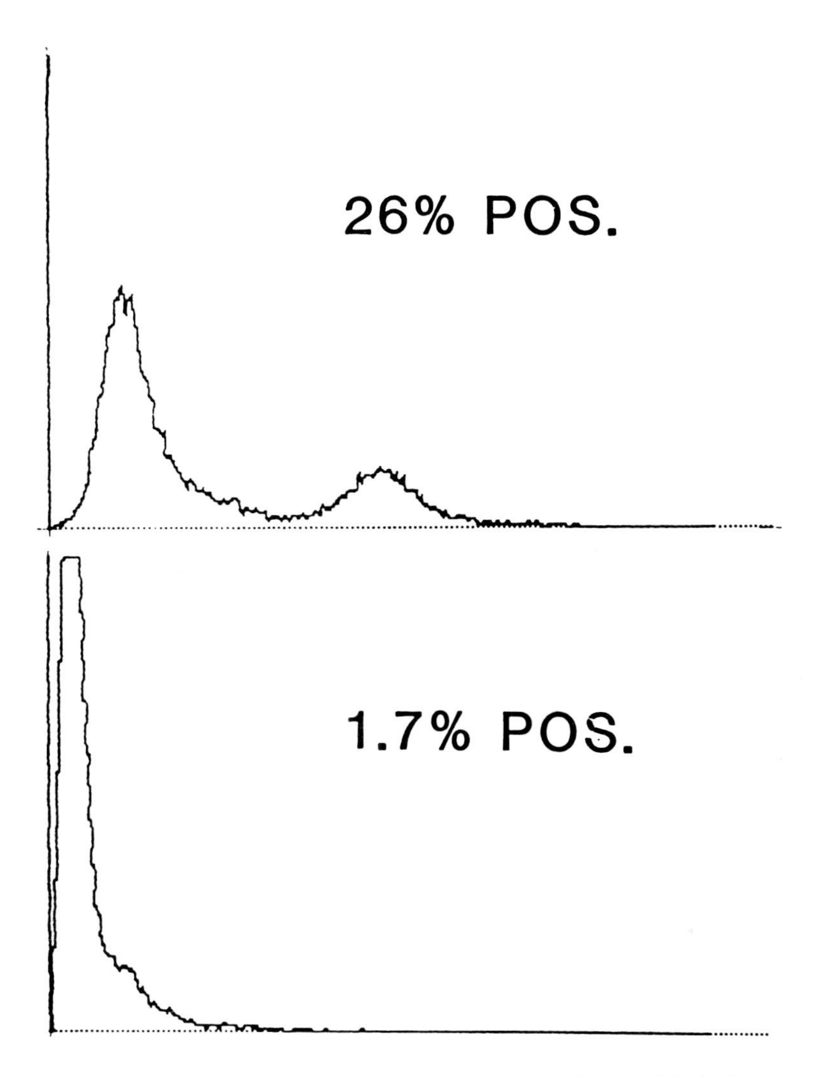


FIGURE 12. Fluorescence histograms of fluorescein diacetate (FDA)-labeled cells in a cytotoxicity test. The top trace shows a mixture of labeled (26%) leukemic cells with normal blood cells. The lower trace shows the effect of treating the cell mixture with monoclonal antibody against the leukemic cells, followed by complement. Cell killing is reflected by the loss of the FDA-positive population.

2. Working solution: 10 $\mu\ell$ FDA stock diluted to 5 m$\ell$ with PBS; use within 30 min.
3. Cells, sera, and complement: as in trypan blue method (previous section).

*b. Procedure*

1. Labeling: centrifuge cells and resuspend in FDA working solution at $10^7$/m$\ell$. Incubate at room temperature for 30 min, with occasional mixing. Wash cells three times in PBS and resuspend in 1 m$\ell$ PBS.
2. Cytotoxicity: aliquot 50 $\mu\ell$ labeled cell suspension per sample tube. Add 50 $\mu\ell$ antibody. Incubate for 30 min in 37°C water bath. Fill tubes with cold PBS and centrifuge. Remove supernatant and add 50 $\mu\ell$ complement. Incubate at 37°C for 30 min. Dilute with PBS, centrifuge, and remove supernatant. Resuspend cells in 20 $\mu\ell$ PBS/glycerol (50:50) and spot onto microscope slide, or resuspend in 200 $\mu\ell$ PBS for flow cytometry.

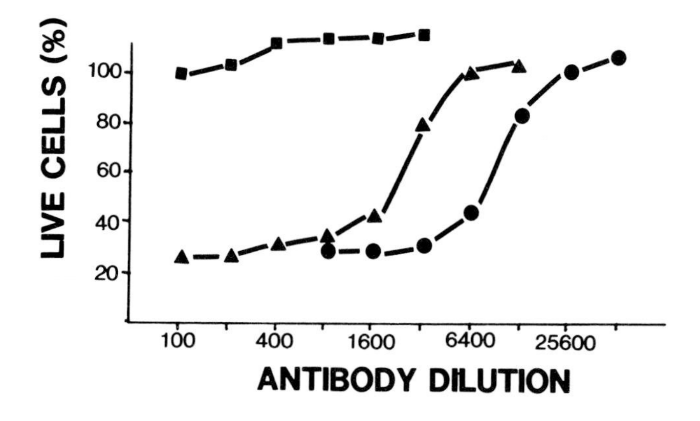


FIGURE 13. Titration curves for one negative and two positive antibodies in a fluorescein diacetate cytotoxicity test, read by flow cytometry (Figure 12).

*c. Notes*

1. This assay may be done in Terasaki microwells, thus, economizing on reagents.
2. Evaluation: fluorescent cells are live; dead cells have lost fluorescence. Examine by fluorescence/phase contrast microscopy and count % cells live, or determine % positive by flow cytometry. Figure 12 shows the FACS histograms obtained from a mixture of labeled (26%) and unlabeled cells, and the same mixture after treating with antibody and complement.
3. See notes on dispensing equipment, complement, and controls in section on trypan blue cytotoxicity (previous section).

*d. Example*

The results in Figure 13 were obtained from an experiment designed to compare the ability of three monoclonal antibodies to kill a population of leukemic cells. One antibody did not kill the cells, while the other two antibodies produced clear titration curves.

*4. Methods for Proliferating Cells*

These methods will not be described in detail, because the cell-killing part of the assay is similar to that described for the other methods, while the methods for determining proliferation include cloning, as described for hybridoma cells in Chapter 3.

Target cells are treated with antibody and complement as described for the fluorescein diacetate method (previous section). Cell numbers may have to be adjusted after calculating the numbers of cells needed for the proliferation assay. All steps must be carried out aseptically, since the cells are to be cultured for several days.

Proliferation of antibody-treated and control cells may be assessed by diluting cells serially into culture wells containing feeder cells. The technique is exactly as described for cloning of hybridomas in Chapter 3, Section IX, except that cells are plated out at numbers ranging from 1 cell per well to $10^6$, in tenfold dilutions. The concentration range used depends on the cell; clearly, a cell line which has a cloning efficiency of 10% will not produce colonies at 1 cell per well. After culture (5 to 15 days depending on the cell type) the numbers of wells showing cell growth at each cell concentration is determined, and comparison of controls with treated cells allows the determination of the percentage of cells killed.

Alternative methods of determining cell proliferation after treatment with antibody and complement involve determining the ability of the cells to incorporate nucleotides. A similar assay depends on the measurement of incorporation of radioactive amino acids[12] rather than cell proliferation.

*5. Chromium Release Method*

Radioactive chromium ($^{51}Cr$, in the form of chromate ion) is taken up by viable cells and binds to protein. It is released from the cell only slowly, unless the membrane is damaged.

*a. Materials*

1. $^{51}Cr$ — sodium chromate — should be in isotonic buffer and high specific activity — for instance, CJS-4 or CJS1P (the latter is sterile and is available in 1 mCi amounts, while CJS-4 is available in multiples of 5 mCi) from the Radiochemical Centre, Amersham.
2. Target cells — at least 90% viable.
3. Medium — RPMI1640 with 10% FCS.
4. V-bottom microtiter tray.
5. Complement — usually rabbit serum.
6. Centrifuge and carrier for centrifuging microtiter trays.

*b. Procedure*

WARNING — $^{51}Cr$ emits a penetrating gamma radiation and should only be handled by staff with adequate training. See Note 3.

1. Cell labeling: prepare cells, $2 \times 10^6$ in 0.2 mℓ medium. Add 0.2 mCi $^{51}Cr$, correcting the volume used for radioactive decay (add 25% more than the nominal amount for each week). Do not use the isotope if it is older than 2 months. Incubate at 37°C on a rocking platform or with occasional gentle mixing in a flat-bottomed tube for 1 hr. Transfer to a conical-bottom 30-mℓ tube and wash the contents of the first tube out with medium, adding the washing to the cell suspension. Wash the cells three times in 20 mℓ medium. Resuspend the cells in 1 mℓ medium, do a cell count, and dilute to $10^6$/mℓ.
2. Check labeling: a 10-μℓ aliquot should give at least 1000 cpm.
3. Test procedure:
   - Add 10 μℓ medium to each well of the microtiter tray.
   - Add 10 μℓ test or control serum to the first well of each row and dilute across.
   - Add 10 μℓ cell suspension to each well, ensuring the cells stay evenly suspended during the whole aliquoting procedure.
   - Incubate 30 min at 37°C in gassed incubator.
   - Add 10 μℓ complement to each well (except no-complement controls).
   - Incubate further 30 min in gassed incubator.
   - Add 100 μℓ cold PBS to each well.
   - Centrifuge plate in refrigerated centrifuge at 4°C.
   - Carefully withdraw 50-μℓ samples for counting in gamma counter.

*c. Notes*

1. The half-life of $^{51}Cr$ is 27 days.
2. The most readily measured radiation emitted by $^{51}Cr$ is gamma radiation, and the radioactivity is, thus, counted in a gamma counter.

3. The gamma emission is of high energy and it is necessary to adequately shield the isotope, particularly the stock vial, both when handling it and during storage. The stock should be kept refrigerated inside a lead container, away from areas where staff spend significant amounts of time and away from photographic material and radioactivity counters. Local regulations concerning disposal of the isotope should be examined before starting work with this isotope.
4. See notes on dispensing equipment, complement, and controls in the section on the trypan blue test (Section 2.c). Because of the potential precision of the Cr release test it is worth using precise volumetric apparatus, and the use of Hamilton repeating dispensers is recommended. It is important to ensure that the cells are kept evenly suspended and do not settle out in the syringe.
5. Precision may be improved by using a different isotope as a volume marker.[13] If $^{14}C$-sorbitol is added to the labeled cells and the ratio of $^{51}Cr$ to $^{14}C$ is measured, any lack of precision in the volume sampled is cancelled out. This procedure significantly complicates the whole experiment, because the isotopes have to be measured in a liquid scintillation counter, and is only recommended if the highest precision is required.
6. Automated harvesting equipment is available, but in the author's experience this is not capable of the precision possible with careful manual harvesting.
7. Results are normally expressed as % Cr release. This requires the determination of the total releasable isotope. This may be done by counting five aliquots of cells dispensed directly into counting tubes at the time when the cells are initially dispensed, and correcting for the volume sampled. However, since lysis of 100% of cells does not necessarily release 100% of the label into the supernatant, an alternative method is to place five additional aliquots in the microtiter tray and add a volume of a detergent solution equal to the volume of serum dilution and complement added to the other wells. These wells can then be treated in exactly the same way as the rest of the plate, and the mean counts for these five wells provide the 100% value for releasable radioactivity. A suitable detergent solution is 1% sodium deoxycholate in water, but any lysis buffer used for making membrane extracts of cells (see Chapter 6, Section IV.C.1.b) may be used. A useful additional control is an antibody known to kill the cells. This serves as a control for complement activity and gives a value for the maximum antibody-mediated Cr release, which may be lower than detergent-mediated release. The detergent control is also needed, however, since a large difference between the detergent-mediated and antibody-mediated isotope release suggests a complement problem.
8. The presentation of results simply as % Cr release, with values given for % release by complement alone, is recommended, since this provides the maximum unfiltered information. However, if it is necessary to transform this into an indication of the % of cells killed the following equation is used:
   where CPM is counts per minute, ab is antibody, C is complement, and max indicates the maximum releasable CPM.

$$\%\ \text{cytotoxicity} = \frac{\text{CPM(ab + C)} - \text{CPM(medium + C)}}{\text{CPM(max)} - \text{CPM(medium + C)}} \times 100$$

9. The titer of a serum is generally expressed as the highest dilution producing >50% lysis.

### G. Dual Marker Methods

There is a number of situations where the coexpression of two or more markers on a single cell needs to be examined. For example, in trying to determine the specificity

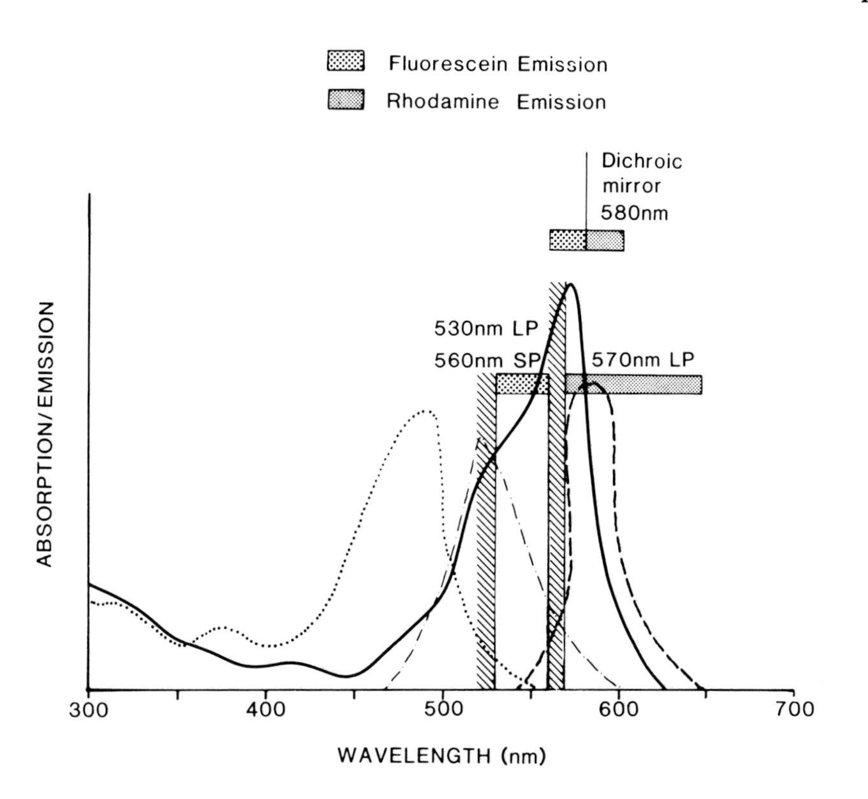


FIGURE 14. Absorption and emission spectra of fluorescein and rhodamine illustrating the method of selecting the optical configuration. An excitation wavelength which stimulates both dyes must be used, and spectral "windows" which reduce "cross-talk" — signals from one dye recorded in the windows intended for the other dye. With these two dyes, however, the overlap of the emission spectra makes some cross-talk unavoidable. Electronic compensation is used to reduce cross-talk, but also reduces signal strength. ······, fluorescein absorption; rhodamine absorption; —·—·— · fluorescein emission; — — — — — rhodamine emission.

of a new monoclonal antibody coexpression with known markers of lineages needs to be determined. Coexpression of two markers may identify a particular subpopulation, for example, the coexpression of T-lineage markers and the major histocompatibility complex class II antigen identifies activated T lymphocytes.

There are several different approaches to the simultaneous identification of two different markers on the same cell, and a number of difficulties. Clearly, if two monoclonal antibodies are used, each labeled with a different fluorescent dye, cells showing both colors coexpress the antigen. This is the simplest solution and works well provided the reagents react strongly enough. The two dyes most often used together for fluorescence microscopy are fluorescein and rhodamine, and the manufacturers of microscopes provide filter systems which allow good transmission of light from each dye while blocking the other. This means that each cell must be looked at under white light and with each of the two fluorescence filter blocks in position. Filter blocks which allow both colors to be seen simultaneously are generally less satisfactory, because it is not easy to determine when both dyes are present together, especially if a particular cell stains brightly with one dye and faintly with the other. The use of fluorescein and rhodamine simultaneously in flow systems has been less satisfactory, because of the difficulty of obtaining good excitation of both dyes with a monochromatic light source, and the overlap of the emission spectra of the two dyes. The excitation and emission spectra of fluorescein and rhodamine are shown in Figure 14.

Flow cytometers equipped with two lasers and an optical system such that the cell traverses the two laser beams in sequence can analyze the data from two dyes with overlapping spectra because the signals are separated by a short time interval. Most instruments have a single light source for fluorescence, and other solutions to the problem of spectral overlap are required. Fluorescein and rhodamine have been successfully analyzed by a variety of maneuvers, including using all-lines excitation (removing the monochromator) so as to excite both dyes effectively, and using narrow-band filters to resolve the emission spectra. Flow cytometers are equipped with electronic devices which compensate for overlap of the emission spectra of two dyes. However, any device which subtracts the overlapping signal, either electronically or by spectral filtering, also subtracts from the signal being measured (see Figure 14), and reduces sensitivity. The development of new dyes has, to some extent, superseded these methods. In particular, the biological dye phycoerythrin pairs well with fluorescein, because both can be excited efficiently at 488 nm and their emission spectra are sufficiently well resolved, using band filters. At present, however, most work has been done with commercially prepared phycoerythrin conjugates.

Two-color direct fluorescence requires the use of labeled monoclonal antibodies. Furthermore, staining intensities are often weak in direct fluorescence. Two-color indirect fluorescence can be carried out provided the second antibodies can distinguish between the two monoclonals (or if one of the primary antibodies is derived from a different species). For example, two monoclonals of different subclasses may be used, provided conjugated subclass-specific second antibodies are available. An alternative approach is to attach the monoclonal antibodies to fluorescent microparticles.[8,14] The two monoclonal antibodies may be derivatized differently; for instance, biotin-labeled antibody detected with fluorescein-avidin and dinitrophenylated (DNP) antibody detected with phycoerythrin-conjugated anti-DNP.

Dual marker methods are not restricted to fluorescence. It is possible to combine rosetting and fluorescence,[15] or membrane fluorescence with cytotoxicity.[16] Simultaneous analysis of DNA content and membrane fluorescence may be carried out by flow cytometry.[17] Combined analysis of monoclonal antibody reactivity, enzyme histochemistry, and morphology can be carried out at light[18] and electron[19] microscope levels, and can provide valuable information on the variability of phenotype within a differentiation lineage.

In some instances, it is possible to derive information on the coexpression of two markers without dual labeling. In assessing whether a new monoclonal antibody stains the same population of cells in blood, the cells may be stained with each antibody separately and with the two together. If the populations stained are not identical the mixture will stain a greater number of cells than either antibody alone. This method works well in some instances, but interpretation is often difficult, particularly if the antibodies stain with different intensities.

## III. TISSUE SECTIONS, SMEARS, AND CYTOCENTRIFUGE PREPARATIONS

The study of cell phenotype *in situ* in tissue sections can provide information which is lost if the tissue is disaggregated. In particular, the topographic relationships between different cell types can tell us something about the way in which the functions of different cells are coordinated. Often a cell cannot be identified purely from its structure and immunological phenotype, but the combination of this information with the location of the cell in tissue allows a positive identification. Furthermore, the methods used to disaggregate tissue may destroy some markers. Even with cells which are normally found in suspension, such as the blood cells, there are reasons for wishing to examine smears or cytocentrifuge preparations. Studies on blood smears are rapid and technically uncomplicated, although they lack the quantitative precision of flow cytometry.

### A. Types of Tissue Preparation

#### *1. Blood Smears*

There is a long tradition of diagnosis of blood diseases by examining blood smears — a drop of blood placed on a glass slide and spread out using the edge of another slide. The smear may be stained to reveal morphological details of the white cells, while the red cells, present in great excess, form an unobtrusive background. The relatively recent use of immunological markers to identify white cell populations and subpopulations has generally required purification of the white cells and has been accompanied by the use of sophisticated equipment, including flow cytometers. However, useful data, adequate for many diagnostic applications, can be obtained by carrying out immunological phenotyping on blood smears and cytocentrifuge preparations. Several procedures have been published.[20,21] Apart from the fact that this technique can be carried out in any good hospital diagnostic laboratory, without the need for sophisticated equipment, the blood-smear method can be particularly useful for the detection of rare cells. If the staining is carried out well, a single positive cell will stand out clearly in several microscope fields of negative cells. However, the wider application of this method is restricted at present because many of the antigens that the hematologist would wish to detect are not stable to fixation. As will be discussed in greater detail below (Section B), stability to fixation is a function of the antigen, the antibody, and the fixative used. Thus, it is very likely that suitable combinations of monoclonal antibodies and fixatives will be developed, in due course, to enable the hematologist to carry out routine analysis of cell phenotype on blood smears.

#### *2. Cytocentrifuge Preparations*

While immunofluorescence is usually the method of choice for the analysis of cell phenotype in suspension, there is a number of situations where an alternative method is preferable. The routine methods for immunofluorescence require large numbers of cells ($5 \times 10^5$ per antibody test). This requirement can be reduced, but not by much. If cells are deposited on a microscope slide in a cytocentrifuge, the number of cells required is $2 \times 10^4$ per slide, and if cells are in short supply $5 \times 10^3$ will do. Cytocentrifuge preparations are also preferable when the internal components of the cell are to be examined rather than just the outer membrane.

Cells which have been deposited on a slide in a cytocentrifuge are best stained by an immuno-enzyme method, rather than immunofluorescence. This carries the advantage that the preparations are semipermanent and can be examined using an ordinary microscope. The preparation can be counterstained for morphological examination, and a variety of cytochemical stains for individual enzymes can be used in combination with antibody-based staining.

Cytocentrifugation requires the use of a special centrifuge, the cytocentrifuge. Cells are suspended in 50% serum (FCS, for example) at a concentration giving $2 \times 10^4$ cells in 50 to 100 $\mu\ell$. The suspension is loaded into the slide carrier and the cells sedimented according to the cytocentrifuge operating instructions (typically, centrifuge at $500 \times G$ for 5 min at room temperature). The slides are removed from the centrifuge and allowed to dry in air for several hours (or overnight) at room temperature. The cells are then fixed, for example, in periodate/lysine/paraformaldehyde (PLP) for 15 min at 4°C. Different fixatives may be needed for different antigens (see Section III.B). Fixed slides may be stored at −80°C wrapped in foil and in a box containing dessicant.

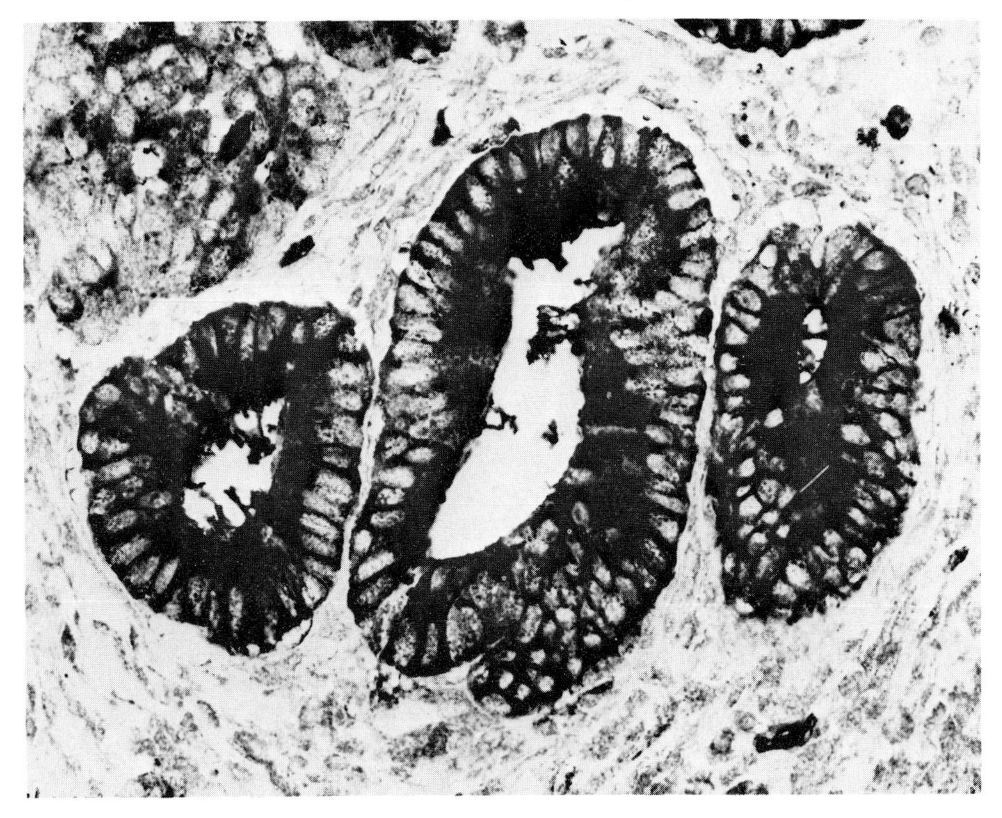

FIGURE 15. Staining of adenocarcinoma of esophagus with the monoclonal antibody FMC61, directed against a human epithelial cell antigen. Paraffin section, 3 to 6 μm, of tissue fixed in buffered formaldehyde. Antibody was detected using an immunoperoxidase procedure (see Section III.E.3) and the section was not counterstained. (Photograph: Dr. A. J. Hodgson.)

### *3. Tissue Sections*

The traditional method for the study of solid tissues involves embedding the tissue in a paraffin or plastic block before cutting sections with a microtome. This method gives excellent preservation of morphology, but the series of steps necessary to fix and dehydrate the tissue before embedding, and to rehydrate it before reaction with antibody, damages many antigens. This problem has led to the development of cryostat sectioning, in which the tissue is frozen, either fixed or unfixed, and sectioned, while still frozen, in a special microtome, the cryomicrotome. Cryostat sections preserve antigenicity better, but morphology is often poor. Some monoclonal antibodies do work very well with paraffin block material. These are particularly useful, because paraffin embedding is still the most widely used procedure, and because the monoclonal antibodies can be tested retrospectively on the large amounts and variety of archival material available in most pathology departments. Figure 15 shows staining of an epithelial antigen by a monoclonal antibody on fixed, embedded tissue.

## B. Fixatives

The ideal fixative for immunohistochemistry preserves morphology and antigenicity (the ability to react with antibody), allows antibody to stain internal components of the cell, and preserves spatial relationships between cell constituents. The ideal fixative for immunohistochemistry does not exist. Fixation procedures which preserve morphology best tend to destroy antigenicity or restrict penetration by antibody, and successful procedures are based on compromise. Such compromises tend to involve reactions which are not allowed to go to completion (for example, brief treatment with low concentrations of cross-linking agents). Because these reactions are not allowed to go to completion, the outcome tends to be variable; slight changes in time, pH, temperature, or concentration of fixative affect the preservation of both morphology and antigenicity.

Even when experimental conditions are controlled rigorously, the differences in chemical nature of the epitopes detected by different antibodies mean that there is no single fixative which preserves antigenicity optimally. For instance, periodate-lysine-paraformaldehyde (PLP) works well with many protein and glycoprotein antigens,[22] but the periodate destroys some carbohydrate epitopes, especially those containing sialic acid.

Fixatives prevent the loss of soluble molecules from the cell, essentially by precipitating them *in situ.* This may result from physical precipitation from solution (by alcohols, for example) or chemical cross-linking (by glutaraldehyde, for instance). Physical precipitation may be reversible or not; if it is reversible, the antigens may diffuse out during staining. As may be expected, precipitation or cross-linking causes reduced accessibility of some epitopes to the antibody. If this happens, staining may be significantly improved by treating the fixed tissue with proteases. These remove some of the precipitated protein and make native protein accessible. This method is likely to be sensitive to minor variations in procedure, since unprecipitated material may diffuse out, and excessive proteolysis may remove the antigen. A detailed discussion of proteolysis may be found in Finley and Petrusz.[23]

Thus, fixation is a complex matter; there is no universal procedure, and the user will need to read the specialized literature (see suggested further reading at the end of the chapter). Technical details will be presented later in this chapter for a few specific examples. Cryostat techniques using unfixed material avoid the uncertainties associated with fixation, but this must be weighed against the deterioration in morphology when using unfixed material, as well as the wider availability of equipment, expertise, and stored material for embedded tissue methods. Furthermore, there is inevitably a greater loss of soluble antigens by diffusion from unfixed cryostat material than from fixed cryostat or embedded tissue. Probably, the best approach to tissue studies is to try first to obtain monoclonal antibodies that work on fixed tissue, and to turn to unfixed tissue only as a last resort.

## C. Staining Methods

### *1. Introduction*

A number of different techniques are used to detect antibody binding to tissue or immobilized cells. Immunofluorescence has been described in detail in the context of cell-suspension studies. There is little to recommend the use of fluorescence for cytocentrifuge or smear preparations. These preparations are used instead of cell suspensions when counterstaining is required to identify the cell types. In tissue sections, on the other hand, immunofluorescence is useful when counterstaining is not required. When conditions are controlled to minimize nonspecific staining, immunofluorescence gives excellent discrimination between stained cells and background.

In the majority of tissue-section studies, and in all studies on cytocentrifuge and blood smear preparations, the need to see morphological features dictates the use of dyes which are visible with white light and which are readily distinguishable from the colors used to counterstain for morphology. Two different types of staining are used. In immunoenzymatic methods an enzyme is coupled to antibody (by a variety of different methods to be discussed below) and a chromogenic substrate is modified by the enzyme to produce the color. The other method relies on particles, especially colloidal gold, attached to the antibody to locate antigen.

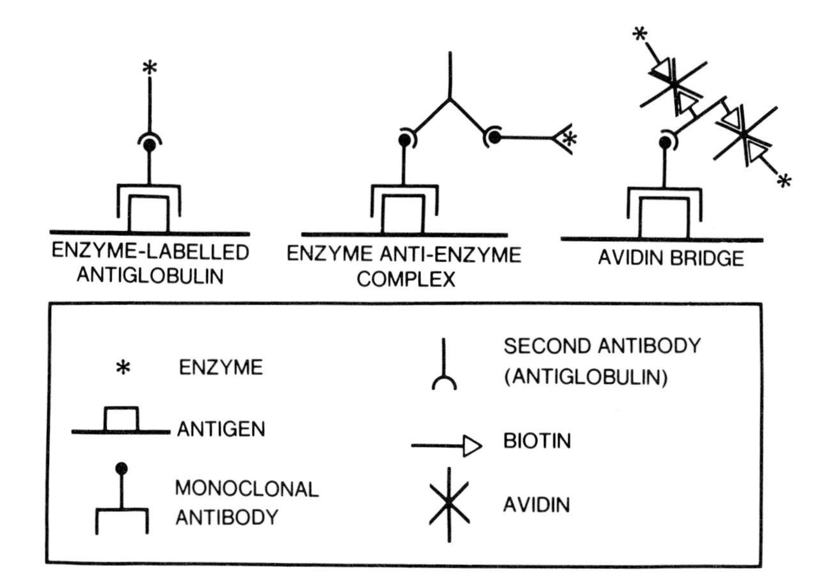


FIGURE 16. Schematic representation of enzyme immunohistochemical staining. See text for detailed discussion.

*2. Immunoenzyme Methods*

The simpler immunoenzyme methods are analogous to direct and indirect immunofluorescence (Figure 2). A monoclonal antibody can be coupled directly to enzyme. This procedure is rarely, if ever, used, although it may be worth considering for dual-marker studies, if the antigen is present in sufficient density to compensate for the relative insensitivity of the direct method. The indirect method involves the coating of the tissue with monoclonal antibody followed by enzyme-conjugated antiimmunoglobulin (Figure 16). The method is technically straightforward, and enzyme-labeled antiglobulins are available commercially (for example, from Sigma, Bio-Rad, and Dakopatts). Sensitivity may be increased by adding extra layers (for example, rabbit-antimouse globulin followed by sheep-antirabbit globulin). Usually, only the last layer is enzyme labeled, but each antiglobulin reagent may be enzyme conjugated, to increase the concentration of enzyme. Each reagent must be checked for nonspecific binding to the target tissue. Nonspecific binding, especially to immunoglobulin of the target species, is often a major problem with this technique. The need for controls and the methods for reducing nonspecific binding, emphasized in the discussion on immunofluorescence (Section II.A.2.d. Note 6), apply equally to immunoenzyme methods which utilize an antiglobulin reagent.

Enzyme-antienzyme complex methods have achieved considerable popularity in recent years, and the availability of mouse monoclonal antibodies against peroxidase and alkaline phosphatase has made this technique attractive for use with mouse monoclonal antibodies. The principle of the enzyme-antienzyme method is illustrated in Figure 16. The target tissue is reacted with monoclonal antibody, leaving it coated with mouse immunoglobulin. A monoclonal antibody against the enzyme is mixed with the enzyme to produce soluble immune complexes (enzyme-antienzyme complex). An antiglobulin reagent (for instance, rabbit antimouse immunoglobulin) is added to the tissue in excess, so that the tissue is coated with antiglobulin, which retains free-binding sites for mouse immunoglobulin. These free-binding sites can then bind the enzyme-antienzyme complex.

The complex may be prepared using mouse monoclonal antienzyme antibody or using polyclonal antibody of a different species and an extra layer of antibody. An example was described by Hancock et al.[24] using mouse monoclonal antibody, rabbit-antimouse immunoglobulin, swine-antirabbit immunoglobulin, and a rabbit peroxidase-antiperoxidase complex. Mouse monoclonal antibodies against horseradish peroxidase and alkaline phosphatase have been described.[25,26] Preformed complexes, with the immunoglobulin deriving from various species, are available commercially (Dakopatts [PAP and APAAP]; Bio-Yeda [PAP]; DP Diagnostics [APAAP]).

The enzyme-antienzyme methods appear to offer greater sensitivity than enzyme-linked antiglobulin methods. This has been attributed to the formation of large enzyme-antibody complexes containing several molecules of enzyme. However, the stoichiometry and size of the complexes have not always been determined, and when they have, the complexes have turned out to consist of 2 molecules of enzyme and 1 of antibody.[26] An alternative explanation for the lower sensitivity of enzyme antibody conjugates is that the coupling process may inactivate a proportion of the enzyme and antibody molecules. Furthermore, the binding of enzyme-conjugated antiimmunoglobulin is inhibited by any unconjugated antiglobulin, if the conjugate has not been adequately purified.

In enzyme-antienzyme methods, the specificity of the bridging antibody is less important than in antiglobulin methods. If the bridging antibody preparation contains contaminating antibody that reacts with an element on the target tissue, this antibody should not bind the complex. The only antibodies which will give nonspecific staining will be genuinely cross-reactive antibodies, which react with a determinant present both on the target tissue and on the antienzyme antibody. Such antibody does occur, since immunoglobulins from different species share some determinants. However, in practice this is much less likely to cause nonspecific staining than the use of the same antiglobulin reagent conjugated to enzyme.

Another variant on the enzyme method uses the high-affinity reaction between biotin and avidin. The monoclonal antibody, or an antibody against mouse immunoglobulin, is labeled with biotin. Avidin binds to the antiglobulin by reacting with the biotin group. The avidin may either be directly conjugated to enzyme or may be previously complexed to biotinylated enzyme (Figure 16). The latter variant is similar in principle to the enzyme-antienzyme method, but utilizes avidin as a bridge and biotin as the common determinant bound by the bridge (Figure 16). The high affinity of the reaction between avidin and biotin ($10^{15}$ $M^{-1}$, compared with $10^{9}$ $M^{-1}$ for an avid antigen-antibody reaction) and the availability of several biotin sites on biotinylated antibody combine to make the avidin-biotin system very sensitive. Avidin can give nonspecific binding, particularly to nuclear proteins. This can be reduced by using a bacterial product with similar biotin-binding properties, streptavidin. The avidin-biotin-enzyme method has become very popular, owing to its high sensitivity and the availability of a good commercial kit (Vectastain, Vector Laboratories). The alternative of conjugating the avidin directly to enzyme is reported to be capable of even greater sensitivity.[27]

A major advantage of the methods which utilize either antibody or avidin as a bridge, with unconjugated or biotinylated enzyme, is that they obviate the need to couple enzyme to antibody. Methods for chemical coupling of enzyme to antibody are not described here, since the conjugates are available commercially (from Sigma, Bio-Rad Laboratories, and from Dakopatts, among others). The coupling methodology is complicated by the need to avoid loss of enzyme and antibody activity. Methods using homo-bifunctional coupling reagents (cross-linking reagents with two identical reactive groups) produce enzyme-enzyme and antibody-antibody polymers, as well as the required enzyme-antibody complexes. On theoretical grounds, therefore, hetero-bifunctional reagents (which are reacted with each of the proteins separately and, in a separate step, form enzyme-antibody conjugates) are preferable. The homo-bifunctional reagents are more commonly used, however, and they do produce material of adequate performance. Methods for coupling enzyme to antibody are reviewed by Boorsma.[28]

A procedure for biotinylation of antibody, which may also be applied to enzyme, is described in Section V.C.

*3. Particulate Markers — Silver-Intensified Gold Labeling*

Rosette procedures have been discussed in the context of cell suspension studies (Section II.E). The use of colloidal gold particles to detect antibody binding has been developed for ultrastructural studies and is described in Section D. The gold particles have been used at the light microscope level on cells in suspension (see Section III.E.6). The sensitivity of the gold method has been increased significantly by using the gold particles as foci for the deposition of silver.[29] The technique is comparable in sensitivity with the best of the enzyme-based methods. Experimental details are described in the example in Section III.E.6. Other examples of the use of this sensitive technique may be found in Somogyi and Hodgson[30] and Hacker et al.[31]

D. Electron Microscopy

Staining with monoclonal antibody at the level of magnification of the transmission electron microscope (EM) may be used to determine the ultrastructural localization of the antigen concerned. Studies at EM level are also useful in obtaining a more definite morphological identification of a cell type reacting with the monoclonal antibody. As in studies with the light microscope, antibody staining may be combined with staining for particular enzymes.

A number of reagents (ferritin, peroxidase) have been used over the years to demonstrate binding of antibody at EM level, but for most purposes the best method is the procedure using colloidal gold as the electron-opaque material. Although there are some difficulties related to the gold-labeled reagents, the method is simple, versatile, and, in spite of the use of a "precious" metal, inexpensive. By using different particle sizes, two or more markers may be demonstrated simultaneously. Furthermore, the labeling of antigen with gold may be combined with enzyme cytochemistry.[19]

The broad range of applications of colloidal gold techniques has been reviewed by Beesley,[32] while a detailed review of technical and theoretical considerations is provided by Roth.[33] Most work has been carried out with indirect methods, utilizing gold-labeled protein A or second antibody (antimouse immunoglobulin). The gold-labeled reagents are available commercially (Jannsen, DP Diagnostics, E-Y Laboratories, Amersham, BRL), but the activity of these reagents is variable, resulting possibly from a short shelf-life. A method for preparing gold-protein conjugates is described in Section V.D. Protein-A gold complexes appear to work more consistently than complexes consisting of antibody, and avidin-gold has been difficult to prepare. Staining of tissue for electron microscopy is carried out on thin sections of tissue. If the cells under study are available in suspension, staining may be carried out either pre- or postembedding. An example of the use of protein-A gold is provided in Section III.E.5.

E. Examples

*1. Introduction*

Because of the many variables in the preparation, fixing, and staining of cells in tissue or deposited on microscope slides, no attempt is made to recommend general methods. The general principles underlying the procedures and the need to select the most appropriate technique for the problem in hand have been emphasized in the earlier parts of this section. Technical details are presented here for a few examples, and

FIGURE 17. Humid chamber used for immunohistochemical staining.

the reader is reminded that for a different antigen, a different fixative or procedure may be better. The preparation of tissue sections is not described here, nor are the technical details of light or electron microscopy. The reader who is not already expert in these areas will need to consult colleagues who are.

*2. Equipment and Materials Common to Several Methods; Stock Solutions*

*a. Equipment*

The following equipment, in addition to general laboratory apparatus, is required for processing of microscope slides:

1. A humid chamber in which slides may be laid horizontally to react with antibody. A large plastic box with a tight-fitting lid may be used, with glass rods to support the slides and moist tissue to keep the atmosphere saturated with water. A suitable arrangement is shown in Figure 17. If the atmosphere is not water saturated, the solutions will evaporate partially, leading to uneven staining and poor morphology.
2. Racks and jars for washing of slides. A good arrangement uses a jar with a magnetic stirrer (Figure 18). A row of jars is used for counterstaining and dehydrating slides (Figure 19).
3. For staining with the peroxidase substrate diaminobenzidine (DAB), a fume cupboard is recommended. DAB is no longer officially classified as a carcinogen, but the fact that it was so classified for a long time means that it is generally treated as a hazardous substance. The staining is carried out in a fume hood, not because of any likelihood of fumes, but because this isolates the work to an identifiable "contaminated" area.

*b. Buffers and Solutions*

1. Tris/PBS: this is similar to PBS-azide used for fluorescence, but is further buffered with 0.01 *M* tris. To make 10 ℓ: commercial tablet or powder for Dulbecco's PBS A to make 10 ℓ (alternatively, see PBS recipe in Chapter 3, Section IV.E). Add: $NaN_3$: 6.5 g (to make final concentration 0.01 *M*); TRIS (base), 12.11 g. Dissolve in distilled water, make to 9 ℓ. Add 1 *M* HCl to bring pH to 7.4. Make to 10 ℓ and check pH.

FIGURE 18. Staining jar on magnetic stirrer used for washing and some stages of immunohistochemical staining.

FIGURE 19. Row of staining jars used for dehydration and counterstaining of slides after immunohistochemical staining.

2. 0.1 *M* Tris buffer, pH 7.6: weigh 12.11 g tris base, dissolve in 900 mℓ distilled water. Adjust pH to 7.6 by adding 1 *M* HCl and bring volume to 1 ℓ.
3. Tris-buffered saline (TBS): 0.05 *M* tris, pH 7.4, containing 8 g/ℓ NaCl. Weigh out 6.05 g tris base, dissolve in 900 mℓ water. Add 8 g NaCl, adjust pH to 7.4 by adding 1 *N* HCl and bring volume up to 1 ℓ.
4. DAB stock solution — WARNING: because of the possible carcinogenic nature of this substance it is handled in the fume cupboard, wearing gloves. All surplus solutions and used apparatus are decontaminated by oxidation using hypochlorite solution (domestic or hospital grade bleach). Empty the contents, without weighing, of a 1-g bottle of DAB (diaminobenzidine tetrahydrochloride, SERVA) into a large (500 mℓ) conical flask. Add 125 mℓ of distilled water which has been made acid by addition of 1 mℓ 10% HCl. Stir for 15 min and add 125 mℓ 0.1 *M* tris buffer, pH 7.4. Aliquot into volumes suitable for single experiments (for example, 20 mℓ) and store frozen at −20°C.
5. 1 *M* Imidazole: dissolve 68.08 g imidazole in 900 mℓ water. Bring pH to 7.4 by addition of 1 *M* HCl, and bring volume to 1 ℓ.
6. Counterstain: Hematoxylin — hematoxylin 1 g (BDH/Gurr); 1 g ammonium alum (aluminium ammonium sulfate); Sodium iodate, 0.2 g. Dissolve in 1 ℓ water with heating and stirring. Add 50 g chloral hydrate and 1 g citric acid. Heat to boiling point and boil for 5 min. Cool and filter.

*3. Example 1: Staining of Cytocentrifuge Preparations for Cytoplasmic Immunoglobulin by an Immunoperoxidase Method*

Cytoplasmic immunoglobulin is detected in cells of the B lineage at certain stages in their differentiation. The fixation procedure must render the cell membrane permeable to the staining reagents and must insolubilize the immunoglobulin, to prevent it from diffusing out of the cell.

*a. Materials*

1. Fixative: the fixative is prepared from 40% formalin solution diluted to 2% in PBS and contains 0.25% saponin. Check the pH and adjust to 7.4 if necessary. The same fixative may be used for flow cytometric analysis of cytoplasmic immunoglobulin, using fluorescent antibody.[34]
2. The antibodies used for this example are monoclonal antibodies against human IgG and IgM. They must be titrated in preliminary experiments; we use ascitic fluid at 1/50 dilution. An excess of antibody will lead to high background staining. We find that sensitivity is improved by using purified monoclonal antibody derivatized with biotin, using the procedure described in Section V.C. The antibody diluent is tris/PBS containing 0.5 to 2% heat-inactivated normal serum of the species used to produce the antimouse antibody, to reduce background staining.
3. Detection reagents: we have used the Vectastain kit (Vector Laboratories). The biotin-conjugated antimouse immunoglobulin must be titrated in a preliminary experiment; a dilution of 1/200 is generally suitable. The avidin-biotin complex is made up 30 min before use, following the manufacturer's instructions for ratios of avidin, biotinylated enzyme, and diluent.
4. Substrate solution: stock solutions have been described in Section 2 above. Imidazole increases the sensitivity of the peroxidase method, probably by interacting with the enzyme to shift the pH optimum. The substrate solution is made up immediately before use (see Procedure).

5. Hydrogen peroxide; 100 vol: the hydrogen peroxide should be taken from a bottle which has been open for less than 6 months; the staining intensity falls off as the hydrogen peroxide ages. This appears not to be due to loss of $H_2O_2$, but to the formation of inhibitory substances.

*b. Procedure*

1. Prepare cytocentrifuge slides (Section III.A.2) with $2 \times 10^4$ cells per slide. Remove the slides from the centrifuge and allow them to dry in air for 2 hr at room temperature. The area with the deposited cells should be ringed with a diamond pen to allow it to be easily found.
2. Fix by immersing the slides in fixative for 10 min. Wash the slides three times in tris/PBS, allowing the slides to soak for 10 min for each wash. If the slides are to be used immediately, they are transferred to a horizontal rack and the antibody applied (see below). If the slides are to be stored, they should be rinsed briefly in water to remove buffer salts, air dried overnight, wrapped in metal foil, and stored at −80°C.
3. Place 1 drop of monoclonal antibody on the appropriate part of the slide and incubate overnight at room temperature in a humid chamber.
4. Wash three times in tris/PBS, at room temperature, with stirring, 10 min per wash.
5. Apply biotin-conjugated antimouse reagent and incubate at room temperature in a humid chamber for 60 min.
6. Wash three times for 10 min in tris/PBS.
7. Add the avidin-biotin-enzyme complex and allow to react for 45 min at room temperature.
8. Wash three times in tris/PBS.
9. Prepare substrate solution; mix in sequence, directly in the staining jar in the fume cupboard: 78 mℓ water, 100 mℓ 0.1 *M* tris buffer, pH 7.6, 20 mℓ stock DAB, and 2 mℓ imidazole 1 *M*, pH 7.4.
10. Place slide rack in this solution, which should be stirred continuously. After 5 min add 100 μℓ of 100 vol hydrogen peroxide. Mix for 15 min, wash in tris/PBS. Alternatively, the entire reaction can be carried out by spotting the reagents onto the slide.
11. Counterstain, dehydrate, and mount. The brown stain produced by DAB is insoluble in aqueous or organic solvents, so that any standard histological stain and mounting medium can be used. Counterstain in hematoxylin (20 sec; adjust to give degree of counterstain required), wash in running water, dehydrate, and mount.

*c. Notes*

Some tissues and cells contain endogenous peroxidase. This may be blocked by treating the specimen in 0.3% $H_2O_2$ in methanol for 30 min after the second antibody step (the biotinylated antibody in the procedure given here). The specimen is washed after the methanol/$H_2O_2$ blocking step. In some situations the presence of sodium azide (0.3%) in the substrate solution will inhibit endogenous peroxidase, but not the added enzyme.

*4. Example 2: Staining of Tissue Sections using Alkaline Phosphatase-Antialkaline Phosphatase*

In this example the staining of human lymph node tissue by monoclonal antibodies directed against cell surface antigens is examined.

*a. Materials*

In addition to the general items listed in Section 2 above, the following are received:

1. Fixative: freshly-prepared, ice-cold buffered formol acetone made up as follows:
   - stock buffer solution — $Na_2HPO_4$, 2 g, $KH_2PO_4$, 10 g, and water, 1 ℓ.
   - working fixative — stock buffer, 10 mℓ, water, 20 mℓ, formalin (37% formaldehyde; BDH), 25 mℓ, and acetone, 45 mℓ.

   The final pH should be approximately 6.6.
2. Monoclonal antibody: the monoclonal antibody should be at a concentration which will saturate binding sites; tissue culture supernatant is often strong enough, but ascitic fluid at dilutions of 1/100 or greater may be used. Higher concentrations are often accompanied by high background staining. Antibody should be diluted in tris-saline containing 0.5 to 2% heat-inactivated serum of the species used to produce the bridging antibody.
3. Bridging antibody (antimouse immunoglobulin): suitable antibodies may be purchased or prepared as described in Section V.A.3 and 4. The antibody is diluted in tris-saline, and the optimal dilution should be determined by experiment.
4. Alkaline phosphatase-antialkaline phosphatase complex. We use a monoclonal antibody, FMC55, prepared in this laboratory, but others are available (see discussion in Section C.2). The complex may be stored for at least 12 months, but should be prepared at least 2 hr before use: alkaline phosphatase 10× stock solution: 100 mg alkaline phosphatase (Sigma Chemical Co., St. Louis, Mo; Cat. #P3877; or equivalent) per milliliter tris-saline. Mix 1 mℓ of stock enzyme solution with 9 mℓ of antienzyme hybridoma culture supernatant. This ratio may need to be varied, depending on the antienzyme hybridoma.
5. Alkaline phosphatase substrate: dissolve 5 mg naphthol AS-MX phosphate (Sigma N4875) in 200 $\mu\ell$ dimethylformamide. Dilute to 10 mℓ with 0.1 *M* tris-HCl, pH 8.2, containing 1 m*M* levamisole. The latter drug is added to block endogenous tissue alkaline phosphatases. Immediately before use add 10 mg Fast Red TR salt (Sigma F1500), mix to dissolve, and pass through a 0.45-$\mu$m membrane filter.

*b. Procedure*

1. Cryostat sections: the preparation and sectioning of tissue is outside the scope of this book; advice should be sought from a regular user.
2. Air dry sections for 2 hr at room temperature. Fix by immersion for 10 min in ice-cold freshly prepared buffered formol acetone.
3. Air dry (for about 30 min) and rehydrate in tris/PBS for 5 min immediately prior to staining. Wipe surplus buffer away and place slide on rack in humid chamber. Without allowing the slide to dry, place 1 drop of monoclonal antibody on section. Leave overnight at room temperature.
4. Wash in tris-saline (three times for 10 min per wash, in a stirred dish).
5. Incubate for 30 min at room temperature with bridging antibody.
6. Wash as above.
7. Incubate with APAAP complex for 30 min at room temperature.
8. Wash as above.
9. Incubate with alkaline phosphatase substrate solution for 15 to 20 min at room temperature. The substrate solution is made up immediately before use. The intensity of staining may be increased by incubating for longer, but nonspecific staining may also increase.

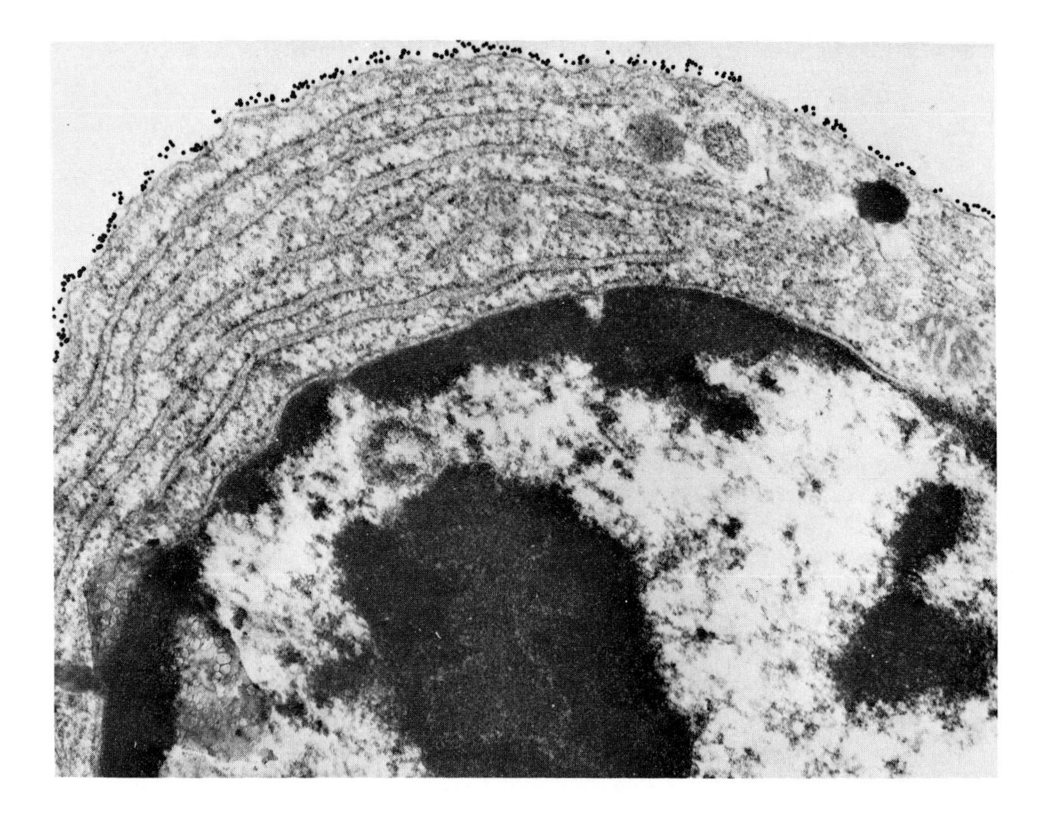

FIGURE 20. Electron micrograph of bone marrow cells stained with the monoclonal antibody FMC34, which reacts with myeloid cells. Staining is seen as gold particles around the membrane. (Photograph: R. J. Hogg.)

10. Wash in tris-saline for 5 min at room temperature, then in water for 5 min.
11. Counterstain and mount: the colored product is insoluble in water, but soluble in organic solvents. Suitable counterstains are Meyer's hematoxylin (20 sec) or methyl green (30 sec). Allow the color to differentiate by washing in running water and mount in aqueous mounting medium (e.g., Glycergel, Dako).

*5. Example 3: Staining at EM Level with Protein-A Gold*

In the example given (derived from studies of Hogg et al.[46]), the expression of surface membrane antigens is analyzed on bone marrow cells, in order to determine the morphological types of cells which do and do not express the antigens. An electron micrograph illustrating gold staining is shown in Figure 20. The approach used is to fix the cells with glutaraldehyde before reacting with monoclonal antibody. Many antigens will not tolerate glutaraldehyde fixation, in which case this step can be postponed until after the reaction with monoclonal antibody. In the example given, the monoclonal antibody is detected with a rabbit-antimouse immunoglobulin followed by protein-A gold. The rabbit-antimouse reagent is used, because the mouse monoclonal antibodies being studied do not all bind protein A, and even if they did the extra layer leads to an increase in sensitivity. The preparation of protein-A gold is described in Section V.D. The protein-A gold may be expected to bind directly to cell-bound immunoglobulin (on B cells or cells which have immunoglobulin bound through Fc receptors) and appropriate controls are, as usual, essential.

*a. Materials*

In addition to the general items listed in Section 2 above, the following are received:

1. Fixative: 2% glutaraldehyde (CEM grade, TAAB) in PBS.
2. Tris-buffered saline containing 0.05% PEG (Carbowax 20 *M*), referred to as TBS/PEG.
3. Monoclonal antibody: should be used at saturation (concentrations used for immunofluorescence should be suitable). Antibody should be diluted in tris-PBS, containing 0.5 to 2% of heat-inactivated rabbit serum.
4. Rabbit-antimouse immunoglobulin (Dakopatts, Cat #Z109); used at 1/50 in tris-PBS.
5. Protein-A gold solution, prepared as described in Section V.D.
6. Phosphate buffer 0.1 *M*; pH 7.4.
7. Osmium tetroxide solution: 1% in the above phosphate buffer. WARNING: osmium tetroxide is a highly toxic vapor. Prepare and use only after obtaining detailed instruction from a regular user.

*b. Procedure*

1. Prepare bone marrow white cell fraction using ficoll/hypaque density layer or other techniques.
2. Fix cells for 15 min at room temperature.
3. Wash once in PBS, then once in TBS/PEG. During these washes and all subsequent steps the cells should be resuspended by gentle pipetting rather than by vortexing.
4. Resuspend cells in TBS/PEG to $10^7$/mℓ. Aliquot for reaction with different antibodies. At least $2.5 \times 10^6$ cells are needed per sample to give an adequate size of pellet for sectioning. Add an equal volume of monoclonal antibody solution. Incubate 60 min on ice.
5. Wash once in TBS/PEG. Resuspend cells in rabbit-antimouse immunoglobulin reagent 200 $\mu\ell/10^6$ cells and incubate on ice for 30 min.
6. Wash twice in TBS/PEG. Resuspend cell pellet in protein-A gold solution, using 10 $\mu\ell/10^6$ cells. Incubate 60 min on ice.
7. Wash once in TBS/PEG and once in PBS. Fix overnight in glutaraldehyde 2% in PBS on ice.
8. Wash twice in 0.01 *M* phosphate buffer, pH 7.4. Resuspend pellet in 1% osmium tetroxide solution in phosphate buffer and stand for 2 hr on ice.
9. Wash twice in phosphate buffer. The pellet is now ready for dehydration, resin embedding, sectioning, and uranyl acetate staining by standard methods (see further reading at the end of the chapter).

*6. Example 4: Staining of Blood Cells with Monoclonal Antibodies for Light Microscopy using Silver-Intensified Protein-A Gold*

Although routine evaluation of lymphocyte markers with monoclonal antibodies is best carried out by immunofluorescence and flow cytometry, there are occasions when, either because the sample is limited or because the flow cytometer is out of action, an alternative technique is required. Particularly when looking for low numbers of cells with a particular phenotype, a method which gives very strong staining with low background is useful. For these reasons we have evaluated the silver-intensified gold method for routine markers studies, as have other workers.[35]

*a. Materials*

1. Specimens: cytocentrifuge preparations of blood cells, fixed in buffered formol acetone (see Section 4).
2. Monoclonal antibodies: undiluted tissue culture supernatants or ascitic fluids previously titrated — a dilution of 1/50 to 1/200 may be a suitable starting level; the antibodies should be used at saturation.
3. Rabbit antimouse immunoglobulin: Dako, Cat #Z109; titrated by prior experiment; 1/50 dilution in tris-PBS containing 1% bovine serum albumin should be suitable.
4. Protein-A gold: see Section V.D. Centrifuge at 8000 × G (microfuge) for 5 min before use.
5. Reagents for silver intensification by Danscher method:[29] gum arabic, crystalline — dissolve 1 kg in 2 ℓ of water by leaving to stand for 5 days with occasional mixing, at room temperature; remove insoluble material by filtration through cotton gauze; store frozen. Citrate buffer: 25.5 g citric acid monohydrate + 23.5 g sodium citrate dihydrate dissolved in 100 mℓ; hydroquinone: 285 mg in 5 mℓ water, dissolve by warming, and make up immediately before use; silver nitrate, 29 mg in 5 mℓ. Make up immediately before use and protect from light. Developer solution (for 24 slides): 20 mℓ gum arabic solution, 3.3 mℓ citrate buffer, 5 mℓ hydroquinone, and 5 mℓ silver nitrate; mix and use immediately.

*b. Procedure*

1. Stain cytocentrifuge preparations with monoclonal antibodies overnight, as described (Section 3).
2. Wash three times in tris PBS (see Section 3).
3. Apply rabbit antimouse immunoglobulin, incubate for 60 min at room temperature.
4. Wash three times.
5. Apply protein-A gold; incubate for 60 min at room temperature.
6. Wash two times in tris PBS, then once in water (because the phosphate ions must be removed before silver is added).
7. Place the slides on a horizontal rack, place a large drop (up to 1 mℓ) of developer on the areas containing the cells, and immediately transfer the slide rack to a dark cupboard.
8. Examine after about 10 min to see if the specimen is darkening. The best development time will vary from run to run, and judging the time improves with experience. Overdevelopment leads to high background staining. A positive and negative control antibody will help to judge the level of development.
9. Wash two times in water; counterstain if required, using hematoxylin; dehydrate and mount (see Section 3.b. listing 11).

## IV. CELL SEPARATION

A monoclonal antibody which has been shown, using the techniques described earlier in this chapter, to react with a membrane molecule restricted to a particular subset of cells can be used to separate that subset from a mixture. A number of different methods are available. These range from flow sorting, which is capable of producing very pure populations on a small scale, to the magnetic bead method, which has been used to process large numbers of bone marrow cells for clinical transplantation. A problem common to all the methods is the evaluation of the efficacy of separation, a problem which is addressed in Section B.

### A. Principles of Separation

The reaction between cells and an immunoabsorbent can proceed in a forward direction (association) and the complex so formed can dissociate to give free cells. The efficiency of separation depends on the equilibrium reached between association and dissociation of cells and immunoabsorbent. This equilibrium depends on several factors, including the equilibrium constant for the antibody-epitope combination, the number and concentration of binding sites on the immunoabsorbent and on the cell membrane, the temperature, time of incubation, and the geometry of the separation system (whether the cells are passed through a bed of immunoabsorbent, allowed to settle on an antibody-coated surface, or caused to interact in some different way). The number of variables affecting separation is, thus, considerable, and it is not possible to devise a method which will always work. Rather, the general principles should be borne in mind when setting up a separation, and the efficacy of separation should be evaluated carefully.

The conditions which are, in general, likely to give good separations will be governed by the nature of the interaction between antigen and antibody. This interaction is an association-dissociation reaction, which reaches an equilibrium. Such reactions, at equilibrium, are described by the law of mass action:

$$K = \frac{[AB]}{[A] \times [B]}$$

where K is the equilibrium association constant, which depends on the energy of interaction between the two reactants (the stronger the interaction between the two, the smaller will be the value for K); [ ] indicates concentration, A and B are two reactants, and AB is the complex. Antigen-antibody reactions are more complex because of the valency of antibody, and the fact that, when dealing with cells, the reaction takes place at a surface, where the effective concentration is very high, and where diffusion is restricted. However, the equation does provide a guide to some of the factors influencing cell separations. The mass action equation tells us that in order to maximize binding, the concentrations of the individual components should be increased as much as possible:

$$[AB] = [A] \times [B] \times K$$

On the other hand, to dissociate the complex the concentrations should be reduced. To maximize binding, a high-affinity antibody (high value of K) should be used. Further general principles, which are not implicit in the law of mass action, but are known from the study of kinetics of chemical reactions, govern the influence of temperature and time. Provided a reaction has reached equilibrium, higher temperatures will generally favor dissociation; lower temperatures will favor association. Thus, binding should generally be done at 4°C, while dissociation will be better at room temperature or 37°C. However, reactions reach equilibrium more rapidly at higher temperatures, and a reaction carried out at low temperature may be suboptimal because equilibrium has not been reached. Another factor involved in reactions at cell surfaces is the fluidity of the membrane. Binding of antibody may depend on movement of the antigen in the membrane to form small aggregates. Membrane mobility is greatly reduced below 15°C.

In conclusion, an understanding of the principles underlying association-dissociation reactions is helpful in formulating working protocols for cell purification, but because of the number of variables, the optimal conditions for a particular separation will need to be found empirically.

Fluorescence-activated flow sorting is less dependent on the variables discussed above than are the other methods. Provided the antibody binds to the cells and this binding can be detected by immunofluorescence, the cells can be separated into positive and negative fractions. Two antibodies which react with the same marker, but with different affinities, can be used in flow sorting and should give essentially identical separations, whereas if the same two antibodies are used in panning or one of the other physical separation methods, the high-affinity antibody may be superior, at least for negative selection (see below).

### B. Assessment of Separation and Optimization of Conditions

The obvious way to assess the efficacy of separation is by analyzing the separated cell populations with monoclonal antibodies of appropriate specificity. However, to obtain an independent assessment of the separation, the monoclonal antibody which has been used for the separation should not be used for the assessment. It is better to use an independent marker, because any deviation from the expected specificity of the antibody, under the conditions of the assay, will not be detected if the same antibody is used both in the separation and in the assay. A second problem with the use of monoclonal antibody to assess the separation is that some of the cells will be coated with monoclonal antibody by the separation process, and this antibody will interfere with assessment by indirect immunofluorescence by binding the antiglobulin reagent. This problem can be avoided by using direct fluorescence or the biotin-avidin system for the assessment, provided that the antibody already on the cell (from the separation process) does not interfere with the binding of the antibody used to assess purity.

These complications hinder experiments intended to optimize the multiple variables which affect separation. In many instances a simple model system may be used to optimize cell separations.[36] A model mixture can be made up by starting with pure populations of cells of the types to be separated, labeling one cell type, mixing them with unlabeled cells of the second type, and then carrying out the separation experiments. For example, cells may be labeled covalently with fluorescein and then followed either by fluorescence microscopy or by flow cytometry. An example is illustrated in Figure 21. The ease of quantitative assessment of the populations makes it realistic to assess the effect of changing the many variables, such as temperature, time, cell concentration, antibody concentration, use of different antibodies, use of one antibody or antibody mixtures, use of a single layer of antibody or a two-layer technique, etc.[36] The limitation of this model system is that it can only readily be applied if the cells are available pure. In some cases cell lines provide the pure populations, but these cells may not behave precisely in the same way as normal equivalent cells. Nevertheless, the model system is useful in optimizing affinity-based cell separations.

### C. Flow Sorting

The analytical use of flow cytometers has been described in Section II.B. Preparative instruments (flow sorters, fluorescence-activated cell sorters) are capable of separating cells into populations which differ with respect to any of the parameters measured. The separation is based on electrostatic charging of droplets containing cells; the charge is positive or negative according to which of two populations defined by the measured parameters (fluorescence, light scatter) the cell belongs to. The charged droplets are separated in an electric field and enter different collecting tubes. The principles are

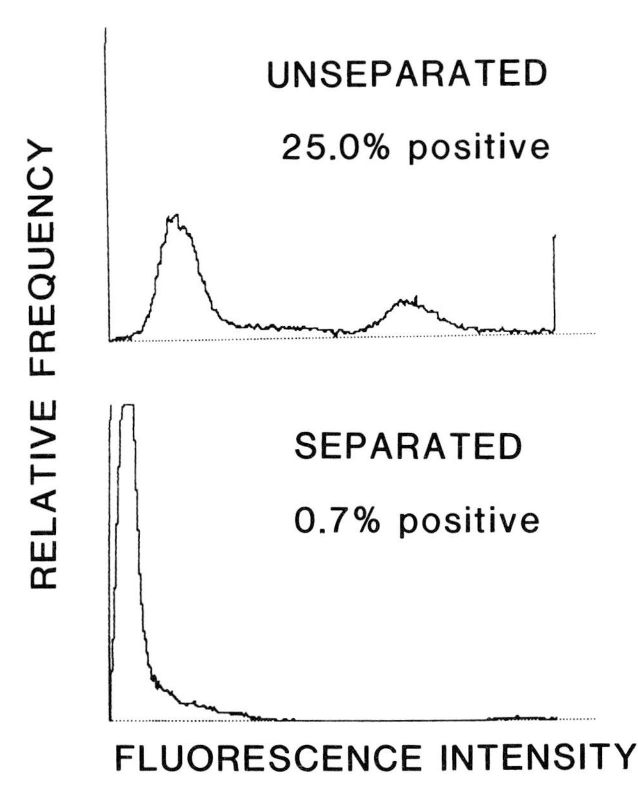


FIGURE 21. Fluorescence histogram of mixture of normal and labeled malignant cells, before (top) and after (bottom) separation using monoclonal antibody-coated magnetic microparticles. See text for details.

illustrated in Figure 22, while Figure 23 shows the ways in which sorting decisions may be made. An example of a cell separation carried out by fluorescence-activated cell sorting with a monoclonal antibody is shown in Figure 24. The sort rate is typically $10^4$ cells per second (input), but for highest purity the sorter must be operated more slowly ($10^3$ cells per second). The rate at which cells can be collected depends on the input rate and on the fraction of the total constituted by the cells being separated. For instance, sorting T cells (70% of blood lymphocytes) from blood is quicker than sorting B cells (5 to 10% of blood lymphocytes). An added factor is the abort rate, which represents cells not selected because the instrument does not have enough information to make a positive "decision". The abort rate is set higher if high purity is required. A detailed description of flow sorting is beyond the scope of this book, and the reader is referred to the instrument operating manuals. Flow sorters are best operated by specialist technicians, who will usually be trained by the instrument supplier.

### D. Immunoaffinity Columns, Panning, Rosetting, and Magnetic Separation

In flow sorting, the antibody reacts with the cell and marks it out for processing in a particular way — drop charging and sorting through an electric field. The antibody is used as a label to identify a cell type, rather than as a handle to isolate the cell. The methods covered in this section use monoclonal antibodies in a more direct way as handles for cell separation. Cells can be separated on affinity columns in much the same way as soluble substances. Alternatively, cells may be attached to the surface of

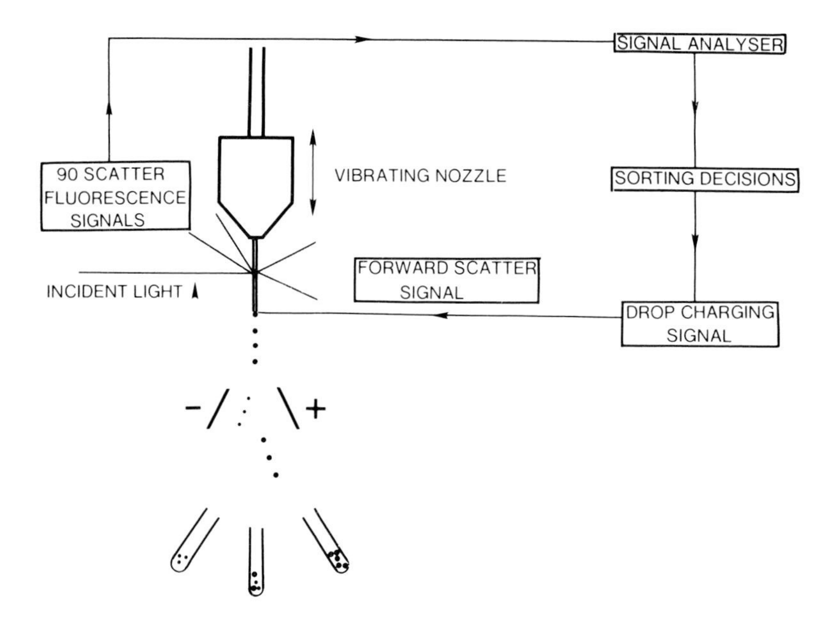


FIGURE 22. Schematic representation of electronic sorting by a fluorescence-activated cell sorter. The stream of cells breaks into droplets, some of which contain cells. The droplets carry either a positive or a negative electric charge, imparted by the instrument on the basis of the measured properties of the cells and the sorting decisions taken by the operator. Sorting decisions are illustrated in Figure 23.

a dish through antibody — the "panning" method first described by Wysocki and Sato.[37] A further approach is to attach particles such as red cells or latex beads to cells through monoclonal antibody, thus, changing the density of cells bearing the antigen and enabling their separation from other cells on the basis of density. If the particles contain magnetic metal, the cells can be separated in a magnetic field.

The choice of method depends partly on personal preference and partly on the desired scale of working. The magnetic bead method was developed for the removal of small numbers of malignant cells from large volumes of bone marrow cells[38] and is particularly suited to this type of application. On the smaller scale generally needed in the laboratory the magnetic method may well also be useful, but has yet to be evaluated extensively. Panning is technically simple (but does not always work straightforwardly) and requires no specialized equipment; it has been used for the separation of subsets of human T lymphocytes[39] and in many other applications. It is our experience that it generally works well for negative selection, but is less efficient at providing pure populations by positive selection. Rosetting is used successfully by laboratories who use rosetting analytically (see Section II.E) and should be particularly effective for positive selection since only cells which have formed stable rosettes will be recovered in the rosetted fraction. Furthermore, the positively selected cells can be recovered, either by lysing the red cells or by gentle agitation of the suspension, which will, in many cases, dissociate the rosettes. In panning, it may be necessary to resort to physically scraping the positively selected cells off the plastic, a process which may cause considerable damage to the cells. Among this group of techniques, the most technologically developed is the use of affinity matrices. Matrices are available commercially ready to couple directly to the antibody, or coupled to antiimmunoglobulin or antifluorescein antibodies.

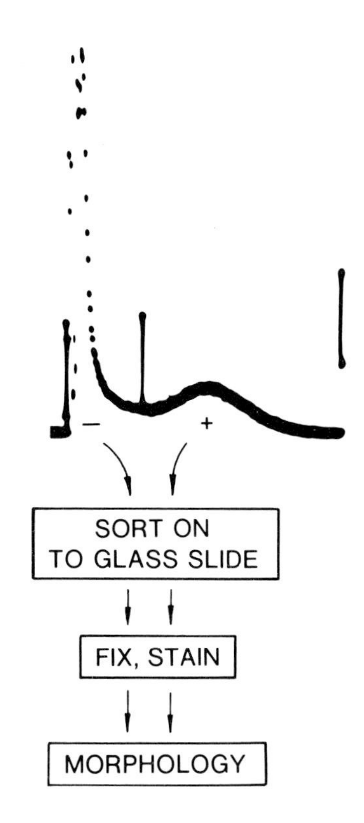


FIGURE 23. Sort decisions: a fluorescence histogram is used to separate cells into fluorescence-positive and fluorescence-negative cells. In the illustration the cells are separated onto slides for morphological examination (see Figure 24); usually they are separated into tubes containing serum to cushion the cells. More complex sort decisions, using up to four parameters, may be made in a manner analogous to the analysis illustrated in Figure 9.

Specific details of separation methods vary widely and, as emphasized earlier, the user should optimize the variables to suit the separation sought and the antibodies available, rather than follow recipes developed by others for different purposes. As a guide, a few examples follow.

### E. Examples of Monoclonal Antibody-Based Cell Separations

#### *1. Separation of Blood Mononuclear Cells from Polymorphs using the Polymorph-Reactive Monoclonal Antibody FMC13*

FMC13 reacts with human blood polymorphs, but does not stain lymphoid cells or monocytes. Buffy coat cells were isolated as described in Section II.A.2.d. Note 1 and red cells lysed with Gey's hemolytic medium (Chapter 3, Section IV.4 and VI.C.3. listing 5). Cells were reacted with FMC13 followed by fluorescein-conjugated second antibody (see Section II.A.2) and separated by flow sorting on a FACS IV instrument (Becton Dickinson, Sunnyvale, Calif.) using standard operating conditions recommended by the instrument manufacturers.

Positive and negative fractions were examined morphologically after making cytocentrifuge preparations and staining with Romanovsky stain. Figure 24 shows the positive and negative fractions. Morphologically, it is clear that the FMC13-positive population consists very largely of polymorphonuclear cells, while the negative population is composed largely of mononuclear cells.

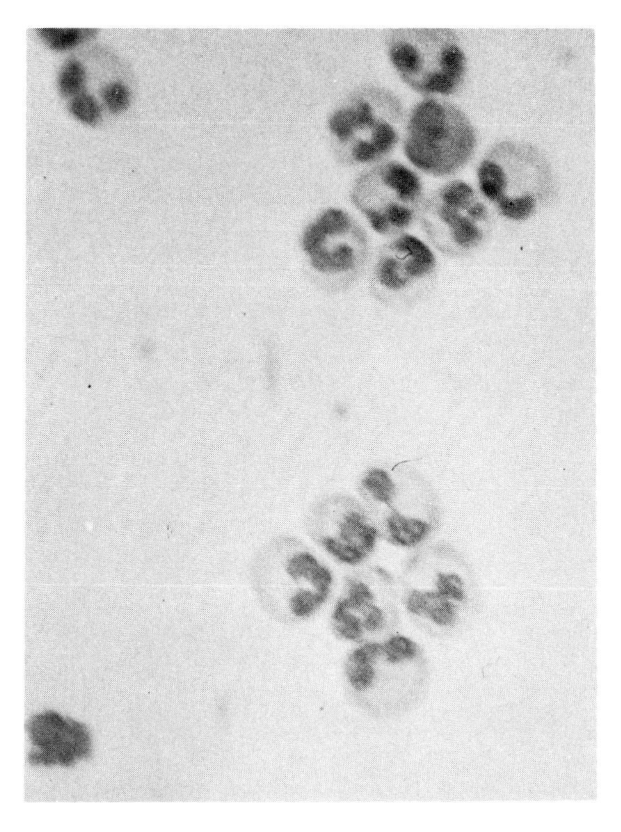

A

FIGURE 24. Cells separated by flow sorting using monoclonal antibody FMC13, which reacts with blood polymorphs. Buffy coat was used, and gated to exclude red cells, but to leave in polymorphs, monocytes, and lymphocytes. The FMC13-positive fraction (A) shows a majority of polymorphonuclear cells, whereas the FMC13-negative fraction (B) consists largely of mononuclear cells (lymphocytes and monocytes).

This separation is, in a sense, trivial, because mononuclear cells may be separated from polymorphs without recourse to monoclonal antibodies or flow sorting. The example is shown because the differences between the cells are readily visible. Flow sorting has been used extensively in the separation of lymphocyte subsets for functional studies. An impressive example of the use of flow sorting has been the enrichment of fetal cells from the maternal circulation, for prenatal genetic screening.[40] This involves the selection of a very small subpopulation (1 in 40,000).

*2. Enrichment of B Cells for HLA-Dr Typing by Panning with Antiimmunoglobulin Antibody*

The major histocompatibility complex class II (MHC class II) antigenic locus is important in sensitizing recipients of organ grafts. The locus codes for at least three polymorphic molecules, DR, DP, and DQ. DR typing is often carried out prior to organ transplantation. The DR antigen is expressed principally on B cells and monocytes, and typing by cytotoxicity is difficult unless these cells are enriched from blood. A variety of procedures are in use in different laboratories. We have used monoclonal antibodies against human IgG and IgM to isolate a fraction rich in B cells.

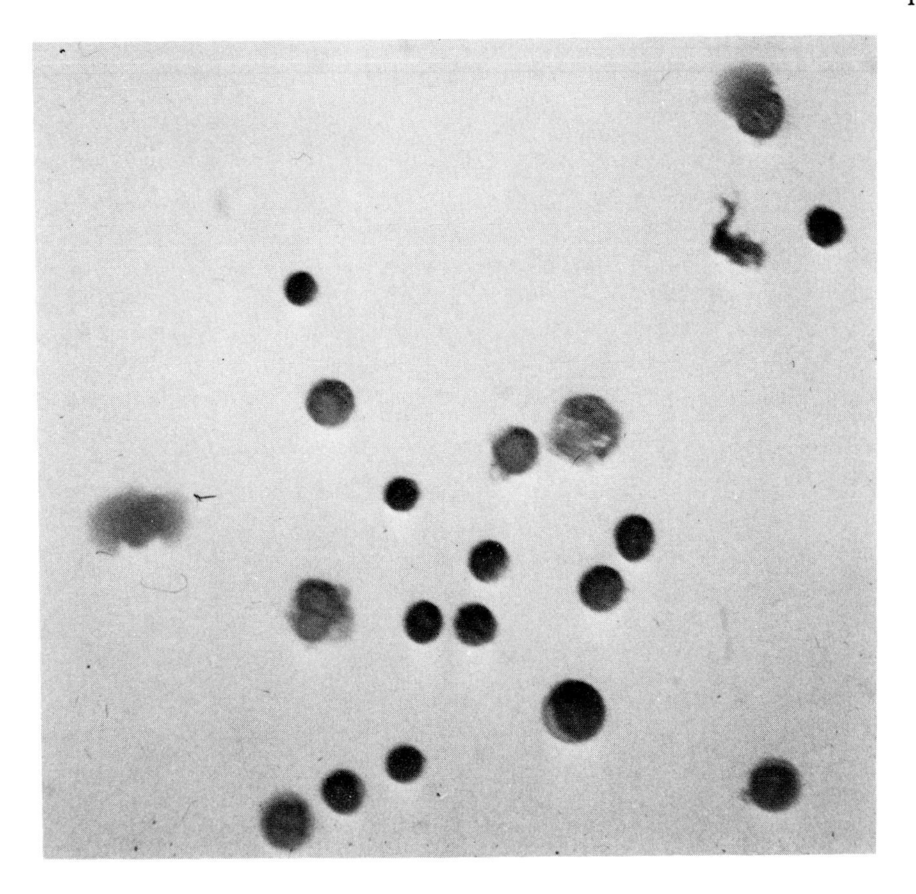

FIGURE 24B

Blood mononuclear cells were prepared by centrifugation over Ficoll-hypaque. The panning plates were prepared by treating petri dishes (50 mm diameter) with monoclonal antibody (3 mℓ 1/50 dilution of ascites fluid; this can be reused at least five times). Plates were incubated with antibody for 30 min at 37°C. Plates were washed and protein-binding sites blocked, by washing the plates three times in warm (37°C) RPMI medium containing 10% FCS.

The cell suspension (3 mℓ at $5 \times 10^6$ cells per milliliter) was placed on the panning plate and allowed to stand for 30 min at 37°C. We generally start with $3 \times 10^7$ cells and use two plates. Nonadherent cells were gently removed, and this was followed by a wash with RPMI medium with 10% FCS to remove any remaining nonadherent cells. The adherent cells were recovered by gently scraping with a rubber "policeman".

The success of the procedure depends very much on technical care taken at all stages. It is particularly important to wash the nonadherent cells off thoroughly but gently, to avoid dislodging adherent cells. The procedure gives cell preparations which are certainly not pure, but sufficiently enriched to allow DR typing. Typical results range from 60 to 80% DR-positive cells, from a starting population with 10 to 30% positive cells. As an alternative to physically dislodging the adherent cells with a "rubber policeman", the cells may be released by incubating the plate in 0.5% EDTA at neutral pH. The success of this method of removal seems to depend on the particular antigen against which the panning antibody reacts.

Monocytes express DR antigens, and T-cell-depleted fractions, consisting largely of B cells and monocytes, may be typed successfully for DR. However, if DQ typing is required, monocytes should be removed before panning. A variety of procedures may

FIGURE 25. Electromagnet in use for separation of cells with antibody-coated microparticles.

be used to remove monocytes. The simplest is to place the cell suspension, in medium containing 10% FCS, on plastic petri dishes for 1 hr at 37°C. Under these conditions monocytes adhere to the plastic. More complete removal of monocytes may be achieved by passing the cells down a small column of Sephadex G10.[41] For the $3 \times 10^7$ cells used for each sample we use a 5-mℓ syringe with a glass-wool plug and 3 mℓ of gel.

Although in this particular example the antibody is applied directly to the plate, better separations may often be obtained in a two-stage procedure, in which the monoclonal antibody is used to coat the cells, while affinity-purified antimouse immunoglobulin is coated onto the panning plate.

### *3. Depletion of Leukemic Lymphocytes from a Blood Cell Mixture by Separation with Monoclonal Antibody on Magnetic Microparticles*

In this example, an artificial mixture of leukemic cells with normal blood cells was prepared, and the leukemic cells subsequently removed by magnetic-antibody particles. The experiment was part of a preliminary study aimed at studying the separation parameters.

#### *a. Materials*

1. Magnet: a series of permanent magnets may be arranged in such a way that a tube can be drawn through the magnetic field, pelleting the magnetic particles at the bottom of the tube. Alternatively, an electromagnet may be used. The magnet used in the author's laboratory is a 1.5-in. Newport Series N38 electromagnet (Oxford Instruments, Oxford, U.K.) and is shown in use in Figure 25. The pole separation is adjustable, allowing use of different tube sizes.
2. Magnetic microparticles: we use ME450 4.5-μm beads[42] kindly supplied by Dr. J. Ugelstad of SINTEF, Trodheim, Norway. These particles are now available commercially (Dynospheres, Norway). Advanced Magnetics (Cambridge, Mass.) supplies a range of magnetic microparticles including particles already coated with antibody against mouse immunoglobulin.

3. Affinity-purified goat-antimouse immunoglobulin: prepared as described in Section V.B.
4. Monoclonal antibody: FMC56, reacting with an antigen expressed on the leukemic cells used in the experiment, but staining <2% of the lymphoid cells.[6]
5. Cells: NALM/6 leukemic cell line and normal blood mononuclear cells isolated on Ficoll-hypaque.

*b. Preparation of Coated Magnetic Beads*

Magnetic beads were suspended in water and vortexed to give a suspension free of aggregates (it may be necessary to use low-power ultrasonic cavitation to disperse aggregates). Particles were counted on a hemocytometer slide. Pilot experiments produced the following data: optimal separation was achieved with a ratio of 100 beads to 1 NALM/6 cell; 1 mg beads (dry weight) contains $1.4 \times 10^7$ beads. Optimal protein/bead ratio was 0.1 mg protein per milligram beads.

Beads and affinity-purified goat-antimouse immunoglobulin were mixed overnight at 4°C, using a rocking platform. Beads were pelleted (using the electromagnet) and resuspended in 1 mℓ PBS containing 1% fetal calf serum. The FCS is used to block any remaining protein-binding sites on the microparticles. The beads were left in the PBS/FCS for 60 min at room temperature and washed once in PBS.

*c. Labeling of Cells*

NALM/6 cells were labeled with fluorescein as follows: fluorescein isothiocyanate (FITC) stock solution — 500 μg/mℓ — this solution may be stored in a dark bottle at 4°C, but will gradually lose activity.

Working-strength solution of FITC is 1/40 dilution of stock solution in PBS.

*d. Procedure*

Cells are resuspended at $10^7$/mℓ in FITC working-strength solution, and incubated for 20 min at 37°C, washed twice in PBS, and finally resuspended in PBS or medium as required.

A mixture was made consisting of 25% fluorescein-labeled NALM/6 cells and 75% unlabeled blood mononuclear cells. The mixture was incubated with FMC56 monoclonal antibody (1/50 dilution of ascites fluid) under the conditions used for indirect immunofluorescence (Section II.A.2). The cells were washed twice to remove excess antibody and resuspended in PBS at a concentration of $5 \times 10^6$ cells per milliliter. The magnetic bead suspension was centrifuged, the supernatant removed, and the beads resuspended directly in the cell suspension. The ratio of beads to NALM/6 cells was 100:1. The mixture was incubated at room temperature for 15 min with occasional mixing, and then gently centrifuged (bring rotor up to speed giving 400 × G and immediately turn off). The sedimented mixture was left on ice for 1 hr, resuspended gently, and the magnetic beads pelleted by drawing the tube slowly through the electromagnet field. The supernatant cell suspension was drawn off for analysis.

*e. Results*

Figure 21 shows the flow cytometric analysis of unseparated and separated cell mixtures. The 25% content of NALM/6 cells was reduced to 0.7% in a single pass.

### F. Preparative Cytotoxicity

*1. Introduction*

The selective killing of one subpopulation is not really a cell separation procedure, but since it serves the same purpose it is described here. Cytotoxicity is an effective

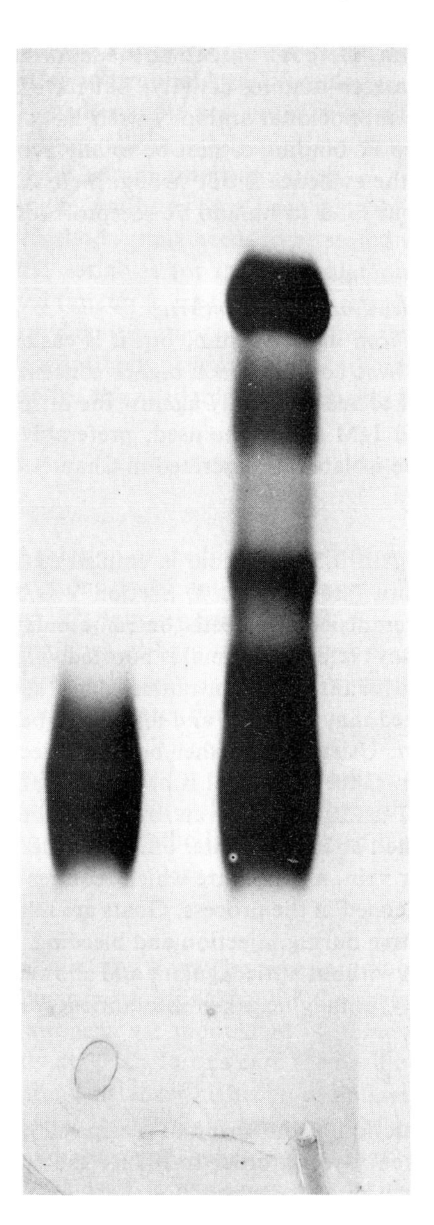

FIGURE 26. Electrophoretic analysis showing purification of the gammaglobulin fraction (left lane) from goat serum (right lane) by octanoic acid precipitation of the nongammaglobulin components.

required for the next process. Protein content and purity may be determined as described in Chapter 4 (Section IV). Figure 26 shows the result of a caprylic acid fractionation of GaMIg.

*5. Fluorescein Conjugation*

The immunoglobulin to be conjugated should be at 10 mg/mℓ and dialyzed against carbonate buffer (0.1 $M\,Na_2CO_3$ in 0.1 $M\,NaCl$), pH 9.8, in the cold overnight. Weigh out fluorescein isothiocyanate (FITC), 30 mg/mg of protein to be labeled. Dissolve in minimal volume of carbonate buffer. Add FITC solution dropwise to protein solution,

3. Affinity-purified goat-antimouse immunoglobulin: prepared as described in Section V.B.
4. Monoclonal antibody: FMC56, reacting with an antigen expressed on the leukemic cells used in the experiment, but staining <2% of the lymphoid cells.[6]
5. Cells: NALM/6 leukemic cell line and normal blood mononuclear cells isolated on Ficoll-hypaque.

*b. Preparation of Coated Magnetic Beads*

Magnetic beads were suspended in water and vortexed to give a suspension free of aggregates (it may be necessary to use low-power ultrasonic cavitation to disperse aggregates). Particles were counted on a hemocytometer slide. Pilot experiments produced the following data: optimal separation was achieved with a ratio of 100 beads to 1 NALM/6 cell; 1 mg beads (dry weight) contains $1.4 \times 10^7$ beads. Optimal protein/bead ratio was 0.1 mg protein per milligram beads.

Beads and affinity-purified goat-antimouse immunoglobulin were mixed overnight at 4°C, using a rocking platform. Beads were pelleted (using the electromagnet) and resuspended in 1 mℓ PBS containing 1% fetal calf serum. The FCS is used to block any remaining protein-binding sites on the microparticles. The beads were left in the PBS/FCS for 60 min at room temperature and washed once in PBS.

*c. Labeling of Cells*

NALM/6 cells were labeled with fluorescein as follows: fluorescein isothiocyanate (FITC) stock solution — 500 μg/mℓ — this solution may be stored in a dark bottle at 4°C, but will gradually lose activity.

Working-strength solution of FITC is 1/40 dilution of stock solution in PBS.

*d. Procedure*

Cells are resuspended at $10^7$/mℓ in FITC working-strength solution, and incubated for 20 min at 37°C, washed twice in PBS, and finally resuspended in PBS or medium as required.

A mixture was made consisting of 25% fluorescein-labeled NALM/6 cells and 75% unlabeled blood mononuclear cells. The mixture was incubated with FMC56 monoclonal antibody (1/50 dilution of ascites fluid) under the conditions used for indirect immunofluorescence (Section II.A.2). The cells were washed twice to remove excess antibody and resuspended in PBS at a concentration of $5 \times 10^6$ cells per milliliter. The magnetic bead suspension was centrifuged, the supernatant removed, and the beads resuspended directly in the cell suspension. The ratio of beads to NALM/6 cells was 100:1. The mixture was incubated at room temperature for 15 min with occasional mixing, and then gently centrifuged (bring rotor up to speed giving 400 × G and immediately turn off). The sedimented mixture was left on ice for 1 hr, resuspended gently, and the magnetic beads pelleted by drawing the tube slowly through the electromagnet field. The supernatant cell suspension was drawn off for analysis.

*e. Results*

Figure 21 shows the flow cytometric analysis of unseparated and separated cell mixtures. The 25% content of NALM/6 cells was reduced to 0.7% in a single pass.

### F. Preparative Cytotoxicity

*1. Introduction*

The selective killing of one subpopulation is not really a cell separation procedure, but since it serves the same purpose it is described here. Cytotoxicity is an effective method when a particular cell type must be removed. Much of our basic understanding of the functional interactions of different lymphoid cell types derives from early experiments of this type; monoclonal antibodies add a new order of specificity. On a different scale, cytotoxicity is used to remove malignant cells or cells which may induce graft-vs.-host disease from bone marrow for human transplantation.

The technical details of cell killing by antibody have been described previously (Section II.F); for preparative use the method simply needs to be scaled up. It is, however, important to optimize experimental variables for the particular purpose. Important variables include the antibodies used (mixtures are generally more effective than single antibodies), the complement, time and temperature of incubation, the number of treatment cycles, and whether the excess antibody is removed before addition of complement (excess antibody can compete for the available complement, thus, reducing cell lysis).

*2. Procedure*

1. Prepare cells in balanced salt solution or medium, at $5 \times 10^6$/mℓ.
2. Add monoclonal antibody or monoclonal antibody mixture. The concentration should be determined by prior titration; see Section II.F.
3. Incubate 30 min at 37°C using a water bath. Centrifuge cells, remove supernatant, and resuspend in complement dilution determined by prior titration.
4. Incubate for 30 min at 37°C, using a water bath.

The cells may now be washed and are ready for use. Dead cells may be removed by centrifuging over Ficoll-hypaque, which will allow dead cells through, but retain viable mononuclear blood cells. Other separation methods may be used if the Ficoll method is not appropriate.

*3. Notes*

The method described is a general one, and it may be necessary to optimize technical details by analytical cytotoxicity experiments using one of the methods described in Section II.F. For example, some antigens are modulated — removed from the cell surface on incubation with antibody at 37°C. In this case the antibody treatment can be carried out on ice. However, in some instances diffusion of antigen in the cell membrane must be permitted to some degree, to allow the formation of aggregates of antibody large enough to effectively fix complement. Incubation at 20°C may be a suitable compromise. If lower temperatures are used, longer times may be needed to allow the reaction to take place. Even when conditions have been optimized on an analytical scale, it is advisable to check the effectiveness of the preparative cytotoxicity, by using appropriate markers.

## V. PREPARATION OF SPECIALIZED REAGENTS FOR USE IN CELL AND TISSUE STAINING AND SEPARATION

### A. Preparation of Goat-Antimouse Immunoglobulin and Fluorescein Conjugate

*1. Introduction*

Fluorescein-conjugated antibodies against mouse immunoglobulin are readily available from several commercial suppliers, and the quality of these reagents has improved greatly, since the demand increased with the widespread use of hybridoma technology. However, some laboratories will have a volume of work that justifies the preparation of homemade reagent. Because of the presence on many cells of receptors for the Fc

terminal of immunoglobulin, there is a case to be made for the use of immunoglobulin fragments which retain antigen-binding activity, but have lost their Fc end ($F[ab^1]_2$ fragments). However, the monoclonal antibody itself has an Fc part, which can also bind to Fc receptors. Since Fc binding cannot be totally avoided; it must be carefully controlled for. Although the evidence is not strong, there is a general impression that antibodies made in the goat bind to human Fc receptors less strongly than do rabbit immunoglobulins.

*2. Preparation of Mouse Immunoglobulin (MIg)*

MIg may be prepared from mouse serum, but it is easier to prepare from pooled hybridoma ascites, since these contain much higher concentrations of MIg. Since the antiserum will be required to react broadly against the different MIg isotypes, a pool including at least IgG and IgM should be used, preferably including all of the IgG subclasses. The MIg can be isolated as described in Chapter 4, Section III.

*3. Immunization of Goat*

Two milligrams of MIg in 0.5 mℓ saline is emulsified with an equal volume of Freund's complete adjuvant (see Chapter 3, Section V.D.3 for notes on the use of Freund's adjuvant). The emulsion is injected intramuscularly in several sites on the rump of the goat. Thirty days later the animal is boosted with a similar dose emulsified in Freund's incomplete adjuvant, again intramuscularly at multiple sites. One week after the booster a test bleed may be taken and tested (see below). If it is satisfactory a larger bleed may be taken. Usually, a further booster injection will be required, approximately a month later. Once an animal is producing satisfactory antibody several bleedings can be taken at 7- to 10-day intervals, and the animal can then be rested. The goat can be kept and boosted at any time later on. Large bleeds of 300 to 600 mℓ blood are taken from the jugular vein, a procedure which requires the availability of a veterinarian or someone experienced in the process. Goats are reliable animals; they may be relied on to be uncooperative during injection and bleeding.

The blood may be taken without anticoagulant and allowed to clot before separation of the serum. Alternatively, it may be taken into anticoagulant and the cells removed by centrifugation.

*4. Processing of GaMIg*

The immunoglobulin fraction of the serum or plasma should be isolated before conjugation with the fluorescent dye, in order to reduce the opportunity for nonspecific staining. There are many published methods for the purification of immunoglobulin. The caprylic acid (octanoic acid) method[43] is easy and rapid, can be used on any scale from 1 mℓ to several liters, and works well for sera from many species (but not mouse serum). The yield is usually better than 80% and the purity of the immunoglobulin is around 80%.

The serum or plasma is warmed to room temperature (octanoic acid solidifies below 15°C). The pH of the serum or plasma is adjusted to 4.8 by dropwise addition of 1 *M* acetic acid. The required amount of octanoic acid is measured out — 6.8 g or 7.4 mℓ/100 mℓ serum. The octanoic acid is added to the serum in a dropwise manner, with vigorous stirring. Typically, for 100 mℓ of serum the octanoic acid is added over a period of 10 min. Stirring is continued for 30 min, and the precipitate is removed by centrifugation (typically 5 to 10,000 × G for 20 min). The centrifuge must be operated above 15°C. If centrifugation does not fully remove the precipitate, the supernatant may be filtered through paper or glass-fiber filters. The pH is readjusted to 7.0 by dropwise addition of 0.5 *M* NaOH, and the material is dialyzed against the buffer

FIGURE 26. Electrophoretic analysis showing purification of the gammaglobulin fraction (left lane) from goat serum (right lane) by octanoic acid precipitation of the nongammaglobulin components.

required for the next process. Protein content and purity may be determined as described in Chapter 4 (Section IV). Figure 26 shows the result of a caprylic acid fractionation of GaMIg.

*5. Fluorescein Conjugation*

The immunoglobulin to be conjugated should be at 10 mg/mℓ and dialyzed against carbonate buffer (0.1 $M\,Na_2CO_3$ in 0.1 $M\,NaCl$), pH 9.8, in the cold overnight. Weigh out fluorescein isothiocyanate (FITC), 30 mg/mg of protein to be labeled. Dissolve in minimal volume of carbonate buffer. Add FITC solution dropwise to protein solution,

with mixing. Allow to mix end-over-end for 1 hr at room temperature. Adjust pH to 7.4 by dropwise addition of 1 *N* HCl. Separate free fluorescein from conjugate on a small disposable desalting column (for example, PD10 column from Pharmacia), using the manufacturer's instructions for running the column.

*6. Quality Control*

*a. Fluorescein/Protein Ratio*

The fluorescein/protein (F/P) ratio is determined by spectrophotometry at 495 and 280 nm. Because fluorescein absorbs at both wavelengths a correction is needed, and the F/P ratio is given by the equation:

$$F/P = \frac{OD495 \times 2.87}{OD_{280} - 0.35 \times OD_{495}}$$

where OD represents the optical density at the stated wavelength.

*b. Staining and Background*

The new conjugate should be titrated and compared with the batch in use. The test should include monoclonal antibodies which give weak staining, as well as strongly positive antibodies. Negative controls should include staining of B lymphocytes and monocytes, which have the greatest tendency to produce nonspecific staining amongst the blood mononuclear cells. If the conjugate is to be used with other cell types, for instance, granulocytes or tissue cells, these should be included in the evaluation of the conjugate.

*7. Absorption*

Antiimmunoglobulin antibodies frequently react across species. Thus, the GaMIg reagent may react with human immunoglobulin, and if it is to be used on lymphoid cells it will stain B cells directly. Absorption may be necessary in many other situations, and the proper use of controls will always alert the user of a reagent to the need for such absorption. Animal sera frequently contain antibodies against parasites and viruses, for instance. These antibodies may be totally unimportant for most applications, but will clearly be detrimental if the antibody is to be used in the detection of parasites or viruses, or for the study of tissues which may contain these organisms.

Absorption should be carried out on insolubilized material. The practice of adding the antigen in solution, to neutralize the antibody, is inappropriately referred to as liquid-phase absorption. It may sometimes render the antiserum specific, but should be used with great care. The result of adding soluble antigen to antibody is the formation of immune complexes. Some of the immune complexes may precipitate and be removed by centrifugation; some complex will remain in solution and bind strongly to Fc receptors on cell surfaces.

Solid-phase absorbents are made as described in Chapter 6, Section V. Some preliminary experimentation will be required to determine how much absorbent is required to remove a particular cross reactivity. Absorption may be carried out either batchwise or on a column. The immunoabsorbent may be regenerated as described in Chapter 6, Section V.

B. Affinity Purification of GaMIg

*1. Rationale*

If the antibody is to be used in a wide variety of applications (for example, if it is to be marketed), it is obviously impossible to absorb with all the possible cross-reacting antigens. In this situation it may be advantageous to purify the antibody (GaMIg) by positive affinity selection, i.e., by absorbing it and then eluting it off a column of MIg. When considering this step it is important to distinguish between two causes of cross reactivity. If the cross reactivity results from the recognition of a shared epitope, affinity purification will not remove it. Thus, if the GaMIg recognizes, among several MIg epitopes, one which is also found on human immunoglobulin, the antibody against this epitope will be retained during affinity purification on MIg. It can be removed, as described above, on a human immunoglobulin column and the antibody will still function, because there are other MIg epitopes not cross reactive with human immunoglobulin.

The cross reactivity may result from the presence in the serum of other antibodies, resulting from previous exposure of the animal to infection, for instance. Such antibodies, by and large, will not react with MIg. Affinity purification should be effective in removing these antibodies.

*2. Method*

There are three essential components required to purify GaMIg on an affinity column:

1. The antibody.
2. The immunoabsorbent column and its operation.
3. An assay.

The GaMIg is prepared as described above (Sections V.A.3 and 4). The immunoabsorbent consists of mouse Ig coupled to a suitable matrix. Mouse Ig can be prepared as described (Section V.A.2), and it is again important to have as many of the classes and subclasses represented as possible. Coupling of the MIg to a solid matrix, absorption, and elution are carried out as described in Chapter 6, Section V. We have found 3 *M* sodium thiocyanate in water effective in elution. The antibody is diluted and dialyzed, and the column washed immediately after elution in order to minimize damage to the antibody or the immunoabsorbent. The yield off the column will be low if it is excessively large in relation to the load of GaMIg applied, and some preliminary experiments are needed to establish the optimal load and column size.

An assay is needed to develop optimal conditions for the affinity purification and to check the yield and purity of the product. Affinity purification carried out "blind", i.e., without an assay, may be successful, but only by chance. GaMIg activity may be determined in an ELISA assay set up as follows.

*3. ELISA for Mouse-Ig-Binding Activity*

Buffers and general procedures used in ELISA assays are described in Chapter 6, Section II.B.

1. Coat ELISA plates with mouse Ig at 50 μg/mℓ in ELISA coating buffer overnight at 4°C.
2. Wash twice in ELISA wash buffer.
3. Apply the antimouse Ig sample to be tested, using a wide range of dilutions in ELISA wash buffer. Dilutions should cover a range of 1 to 1000 in order to maximize the chances of obtaining a result in the readable range.
4. Incubate at 37°C for 1 hr followed by 2 hr at 4°C.
5. Wash five times in ELISA wash buffer.
6. Apply alkaline phosphatase-antialkaline phosphatase complex, which will bind

to free anti-MIg-binding sites. The alkaline phosphatase is then detected using *p*-nitrophenyl phosphate substrate. The preparation and use of the enzyme-antibody complex and substrate for ELISA are described in Chapter 6, Section II.C.

### C. Biotinylation of GaMIg

#### *1. Rationale*

Monoclonal antibodies may be modified by substitution with structures which allow the use of additional probes. For example, an antibody may be dinitrophenylated (DNP) and, subsequently, detected by anti-DNP antibody. The most successful modification of this type has been biotinylation. Biotin is a natural product which reacts with the egg-white protein avidin. Biotinylated antibody may be detected using avidin substituted with fluorescein or other dyes. Alternatively, the avidin may be directly conjugated to enzyme or used to bind biotinylated enzyme, in variations on the immunoenzyme methods (Section III.C.2).

Substituted antibodies are useful in two-color fluorescence, with avidin-fluorescein providing one color and directly conjugated phycoerythrin-monoclonal antibody as the other color. Biotinylation is also useful when an antiimmunoglobulin reagent would give false-positive results because of the presence of endogenous immunoglobulin. For example, when making or using monoclonal auto- or allo-antibodies, or for studies in mice using murine monoclonal antibodies against infectious organisms, the binding of mouse immunoglobulin to mouse tissue must be detected. The presence of B cells or serum immunoglobulin renders the use of GaMIg to detect the monoclonal antibody difficult. The monoclonal antibody may be labeled directly with fluorochrome, or may be biotinylated and detected with avidin-fluorochrome. The latter procedure is more sensitive. Avidin and various conjugates (to enzymes and fluorescent dyes) are available from several suppliers, including Vector Laboratories and Becton Dickinson.

#### *2. Materials*

1. Biotin-*N*-hydroxysuccinimide ester (Sigma, Calbiochem-Behring).
2. Monoclonal antibody at 1 mg/m*ℓ* in 0.1 *M* borate or carbonate buffer, pH 8 to 8.5. Amines, such as tris, compete with the protein for biotin succinimide ester and should be avoided. The antibody should be dialyzed against the borate or carbonate buffer to remove competing amines.

#### *3. Procedure*

1. Add 2 mg biotin succinimide ester, dissolved in 100 $\mu\ell$ dimethylsulfoxide, to 1 m*ℓ* of antibody solution in a bath of melting ice. The reaction is strongly exothermic, and the ester should be added dropwise with gentle mixing to dissipate the heat. Leave the mixture on ice for 10 min with occasional mixing.
2. Separate the derivatized antibody from free biotin using a disposable desalting column (such as PD10, from Pharmacia).

#### *4. Testing*

The activity of the antibody should be compared before and after biotinylation, to ensure that activity has not been lost. This may be done by any assay used for the antibody in question. The ability of the antibody to bind avidin may be conveniently determined by titrating the antibody in the assay for which it is intended, or by dot-blotting (Chapter 6, Section II.C). Finally, it may be desirable to determine if any unconjugated antibody remains, since this would compete with conjugated antibody in the test. This can be done using an affinity column of insolubilized avidin to remove biotinylated antibody and determining the amount of antibody which passes through the column.

### D. Preparation of Colloidal Gold-Protein A

#### *1. Introduction*

Although the use of colloidal gold as a marker in immunohistochemistry dates from 1971,[44] the technique is only beginning to achieve widespread popularity. As a consequence, the availability of commercial reagents for this technique is limited. Furthermore, the reagents available commercially have been variable in performance, perhaps, because of a short shelf-life. For this reason it is worth considering making the reagents in the laboratory.

#### *2. Materials*

1. Conical flask and reflux condenser: the glassware must be soaked with chromic acid before the first use. After soaking in chromic acid for 24 hr, the flask and condenser are washed with copious amounts of distilled water, dried, siliconized, and washed once more with distilled water. Provided the glassware is reserved for use in colloidal gold preparation, it does not need to be soaked in chromic acid between uses, but should be washed in distilled water.
2. Magnetic stirrer/hot plate and magnetic "flea".
3. 0.22-$\mu$m disposable filters.
4. Narrow-range pH paper.
5. Microtiter tray and microliter dispensers.
6. Sodium citrate 1% in water.
7. Potassium carbonate 0.2 *M* in water.
8. NaCl 10% in water.
9. Tris-buffered saline (TBS) (see Section III.E.2).
10. Polyethylene glycol (Carbowax 20 *M*) 1% and 0.05% in TBS.
11. Gold chloride (tetrachloro-auric acid $HAuCl_4 \cdot 3H_2O$, Merck).
12. Protein A (Pharmacia, Uppsala, Sweden). This should be in a low ionic strength solution (<5 m*M*) and may conveniently be dissolved in distilled water, to 1 mg/mℓ.

#### *3. Procedure*[45]

Note that all solutions should be filtered through 0.22-$\mu$m membrane filters before use, and preferably filtered directly into the reaction flask to avoid picking up particulate material. The gold chloride is best dissolved directly from the ampule without weighing, because it absorbs water rapidly. Clean the outside of the ampule after removing the label, and break the ampule in a beaker. Add 25 mℓ water (for 1 g gold chloride) to dissolve the gold, filter, and store in a dark bottle at 4°C. Avoid contact between the gold solution and metal surfaces.

1. Filter 100 mℓ of distilled water into the conical flask and bring to the boil. Add 0.25 mℓ of gold chloride solution and bring back to boil. When the gold solution is boiling, lift the condenser from the flask momentarily and inject 4 mℓ of citric acid solution into the boiling liquid. The citric acid solution should be added rapidly from a plunger-type dispenser. Boil the mixture under reflux and with stirring for 30 min. The color turns faint blue, then wine-red, as the gold is reduced. Allow to cool.

2. Place a number of 100 μℓ aliquots of the reduced gold solution into wells of the microtiter tray and add different amounts of potassium carbonate (0.001 *M*) to determine how much base is needed to bring the pH to 5.9. This must be done as accurately as possible with narrow range (pH 5 to 7 in 0.1 or 0.2 pH unit steps) pH paper, as the gold solution is not compatible with pH electrodes. Having determined the amount of $K_2CO_3$ needed, add the appropriate amount to the bulk reduced gold (the amount needed will be around 1.5 μℓ/mℓ gold sol). Check the pH of the gold and carefully add more $K_2CO_3$ if needed.
3. Salt will cause the gold to aggregate, and protein coating of the gold will prevent this. The amount of protein to be added is determined by titration in the microtiter wells, by adding 200 μℓ gold sol to each well, adding different amounts of a diluted (1/10) protein A solution to each well, and finally adding 20 μℓ NaCl solution to each well. If the gold aggregates, the solution turns blue; if it does not it remains red. To obtain an accurate estimate the amount of protein A is arrived at by successive approximations; use aliquots differing in steps of 5 μℓ first, then use 1-μℓ steps to arrive at a more accurate volume. Having calculated the amount of protein A gold to be added to the gold sol, add a 10% excess and check that an aliquot of the bulk protein A-gold mixture does not precipitate when 20 μℓ NaCl solution is added to 200 μℓ gold sol. A typical result is that 500 μℓ of protein A solution at 1 mg/mℓ is added to 100 mℓ gold solution.
4. Having added the calculated amount of protein to the bulk gold sol and checked that it does not precipitate on adding salt, mix well and add 1% PEG solution, 1 vol/20 vol gold sol. This stabilizes the coated gold particles further.
5. Wash twice with tris-buffered saline containing 0.05% PEG (TBS/PEG), ultracentrifuging at 85,000 × G for 30 min to sediment the colloidal particles. Resuspend in TBS/PEG, add an equal volume of glycerol (to prevent freezing), and store at −20°C. The colloid is best stored at high concentration and aliquots diluted immediately before use.

### *4. Notes*

1. As described, the method should give 17-nm (diameter) particles. The particle size may be increased by adding less citric acid. Smaller particle sizes may be made using different reducing agents.[33]
2. Similar methods are used at different pH for other proteins, including immunoglobulin.[33] However, the method gives variable results with different proteins, because of differences in overall charge.
3. For a more detailed discussion of the preparation of protein-gold colloids see Roth.[33]

## REFERENCES


1. Hoffman, R. A., Kung, P. C., Hansen, W. P., and Goldstein, G., Simple and rapid measurement of human T lymphocytes and their subclasses in peripheral blood, *Proc. Natl. Acad. Sci. U.S.A.*, 77, 4914, 1980.
2. Johnson, G. D. and Nogueira Araujo, G. M. de C., A simple method of reducing the fading of immunofluorescence during microscopy, *J. Immunol. Methods*, 43, 349, 1981.
3. Gilo, H. and Sedat, J. W., Fluorescence microscopy: reduced photobleaching of rhodamine and fluorescein protein conjugates by n-propyl gallate, *Science*, 217, 1252, 1982.
4. Lanier, L. L. and Warner, N. L., Paraformaldehyde fixation of hematopoietic cells for quantitative flow cytometry (FACS) analysis, *J. Immunol. Methods*, 47, 25, 1981.

5. Gadd, S. J. and Ashman, L. K., Binding of mouse monoclonal antibodies to human leukaemic cells via the Fc receptor: a possible source of false positive reaction in specificity screening, *Clin. Exp. Immunol.*, 54, 811, 1983.
6. Zola, H., Moore, H. A., McNamara, P. J., Hunter, I. K., Bradley, J., Brooks, D. A., Gorman, D. J., and Berndt, M. C., A membrane protein antigen of platelets and non-T ALL, *Dis. Markers*, 2, 399, 1984.
7. Zola, H., Moore, H. A., Hunter, I. K., and Bradley, J., Analysis of chemical and biochemical properties of membrane molecules in situ by analytical flow cytometry with monoclonal antibodies, *J. Immunol. Methods*, 74, 65, 1984.
8. Ghetie, V., Nilsson, K., and Sjoquist, J., Identification of cell surface immunoglobulin markers by protein A-containing fluorescent staphylococci, *Scand. J. Immunol.*, 3, 397, 1974.
9. Uchanska-Ziegler, B., Wernet, P., and Ziegler, A., A single-step bacterial binding assay for the classification of cell types with surface antigen-directed monoclonal antibodies, *Br. J. Haematol.*, 52, 155, 1982.
10. De Waele, M., De Mey, J., Moeremans, M., De Brabander, M., and Van Camp, B., Immunogold staining method for the light microscopic detection of leukocyte cell surface antigens with monoclonal antibodies. Its application to the enumeration of lymphocyte subpopulations, *J. Histochem. Cytochem.*, 31, 376, 1983.
11. Wybran, J., Rosenberg, J., and Romasco, F., Immunogold staining: an alternative method for lymphocyte subset enumeration. Comparison with immunofluorescence microscopy and flow cytometry, *J. Immunol. Methods*, 76, 229, 1985.
12. Cleveland, P. H., McKhann, C. F., Johnson, K., and Nelson, S., A microassay for humoral cytotoxicity demonstrating sublethal effects of antibody, *Int. J. Cancer*, 14, 417, 1974.
13. Zola, H. and Palmer, R. M. J., A double-isotope technique for cytotoxicity assays, *J. Immunol. Methods*, 6, 133, 1974.
14. Higgins, T. J., O'Neill, H. C., and Parish, C. R., A sensitive and quantitative fluorescence assay for cell surface antigens, *J. Immunol. Methods*, 47, 275, 1981.
15. King, M. A., Simultaneous detection of two cell surface antigens by a red blood cell rosette-microsphere binding method, and its application to the study of multiple myeloma, *J. Immunol. Methods*, 72, 481, 1984.
16. Loken, M. R. and Stall, A. M., Flow cytometry as an analytical and preparative tool in immunology, *J. Immunol. Methods*, 50, 85, 1982.
17. Braylan, R. C., Benson, N. A., Nourse, V., and Kruth, H., Correlated analysis of cellular DNA, membrane antigens and light scatter of human lymphoid cells, *Cytometry*, 2, 337, 1982.
18. Mirro, J. and Stass, S. A., Fluorescent microsphere detection of surface antigens and simultaneous cytochemistries in individual hematopoietic cells, *Am. J. Clin. Pathol.*, 83, 7, 1985.
19. Polli, N., Zola, H., and Catovsky, D., Characterization by ultrastructural cytochemistry of normal and leukemic myeloid cells reacting with monoclonal antibodies, *Am. J. Clin. Pathol.*, 82, 389, 1984.
20. Moir, D. J., Ghosh, A. K., Abdulaziz, Z., Knight, P. M., and Mason, D. Y., Immunoenzymatic staining of haematological samples with monoclonal antibodies, *Br. J. Haematol.*, 55, 395, 1983.
21. Lowenthal, R. M., Pralle, H., and Matter, H. P., A sensitive method for immunophenotyping stored leukemia and lymphoma cells with preservation of morphological detail, *Pathology*, 17, 481, 1985.
22. Hancock, W. W., Becker, G. J., and Atkins, R. C., A comparison of fixatives and immunohistochemical techniques for use with monoclonal antibodies to cell surface antigens, *Am. J. Clin. Pathol.*, 78, 825, 1982.
23. Finley, J. C. W. and Petrusz, P., The use of proteolytic enzymes for improved localization of tissue antigens with immunocytochemistry, in *Techniques in Immunocytochemistry*, Vol. 1, Bullock, G. R. and Petrusz, P., Eds., Academic Press, London. 1982, 239.
24. Hancock, W. W., Zola, H., and Atkins, R. C., Antigenic heterogeneity of human mononuclear phagocytes: immunohistological analysis using monoclonal antibodies, *Blood*, 62, 1271, 1983.
25. Mason, D. Y., Cordell, J. L., Abdulaziz, Z., Naiem, M., and Bordenave, G., Preparation of peroxidase: antiperoxidase (PAP) complexes for immunohistological labeling of monoclonal antibodies, *J. Histochem. Cytochem.*, 30, 1114, 1982.
26. Cordell, J. L., Falini, B., Erber, W. N., Ghosh, A. K., Abdulaziz, Z., MacDonald, S., Pulford, K. A. F., Stein, H., and Mason, D. Y., Immunoenzymatic labeling of monoclonal antibodies using immune complexes of alkaline phosphatase and monoclonal anti-alkaline phosphatase (APAAP Complexes), *J. Histochem. Cytochem.*, 32, 219, 1984.
27. Giorno, R., A comparison of two immunoperoxidase staining methods based on the avidin-biotin interaction, *Diagn. Immunol.*, 2, 161, 1984.
28. Boorsma, D. M., Conjugation methods and biotin-avidin systems, in *Techniques on Immunocytochemistry*, Vol. 2, Bullock, G. R. and Petrusz, P., Eds., Academic Press, London, 1983, 155.

29. Danscher, G. and Rytter Norgaard, J. O., Light microscopic visualization of colloidal gold on resin-embedded tissue, *J. Histochem. Cytochem.*, 31, 1394, 1983.
30. Somogyi, P. and Hodgson, A. J., Antisera to gamma-aminobutyric acid. III. Demonstration of GABA in golgi-impregnated neurons and in conventional electron microscopic sections of cat striate cortex, *J. Histochem. Cytochem.*, 33, 249, 1985.
31. Hacker, G. W., Springall, D. R., Van Noorden, S., Bishop, A. E., Grimelius, L., and Polak, J. M., The immunogold-silver staining method. A powerful tool in histopathology, *Virchows Arch. (Pathol. Anat.)*, 406, 449, 1985.
32. Beesley, J. E., Colloidal golds: a new revolution in marking cytochemistry, *Proc. R. Microsc. Soc.*, 20, 187, 1985.
33. Roth, J., The colloidal gold marker system for light and electron microscopic cytochemistry, in *Techniques in Immunocytochemistry*, Vol. 2, Bullock, G. R. and Petrusz, P., Eds., Academic Press, London, 1983, 218.
34. Garner, J. G., Colvin, R. B., and Schooley, R. T., A flow cytometric technique for quantitation of B-cell activation, *J. Immunol. Methods*, 67, 37, 1984.
35. Romasco, F., Rosenberg, J., and Wybran, J., An immunogold silver staining method for the light microscopic analysis of blood lymphocyte subsets with monoclonal antibodies, *Am. J. Clin. Pathol.*, 84, 307, 1985.
36. Zola, H., Krishnan, R., and Bradley, J., A simple technique for evaluation of methods of cell separation, *J. Immunol. Methods*, 76, 383, 1985.
37. Wysocki, L. J. and Sato, V. L., "Panning" for lymphocytes: a method for cell selection, *Proc. Natl. Acad. Sci. U.S.A.*, 75, 2844, 1978.
38. Treleaven, J. G., Ugelstad, J., Philips, T., Gibson, F. M., Rembaum, A., Caine, G. D., and Kemshead, J. T., Removal of neuroblastoma cell from bone marrow with monoclonal antibodies conjugated to magnetic microspheres, *Lancet*, i, 70, 1984.
39. Engelman, E. G., Benike, C. J., Grumet, F. C., and Evans, R. L., Activation of human T lymphocyte subsets: helper and suppressor/cytotoxic T cells recognize and respond to distinct histocompatibility antigens, *J. Immunol.*, 127, 2124, 1981.
40. Herzenberg, L. A., Bianchi, D. W., Schroder, J., Cann, H. M., and Iverson, G. M., Fetal cells in the blood of pregnant women: detection and enrichment by fluorescence-activated cell sorting, *Proc. Natl. Acad. Sci. U.S.A.*, 76, 1453, 1979.
41. Mishell, B. B. and Shiigi, S. M., *Selected Methods in Cellular Immunology*, W. H. Freeman, San Francisco, 1980, 175.
42. Lea, T., Vartdal, F., Davies, C., and Ugelstad, J., Magnetic monosized polymer particles for fast and specific fractionation of human mononuclear cells, *Scand. J. Immunol.*, 22, 207, 1985.
43. Steinbuch, M. and Audran, R., The isolation of IgG from mammalian sera with the aid of caprylic acid, *Arch. Biochem. Biophys.*, 134, 279, 1969.
44. Faulk, W. P. and Taylor, G. M., An immunocolloid method for the electron microscope, *Immunochemistry*, 8, 1081, 1971.
45. Frens, G., Controlled nucleation for the regulation of the particle size in monodisperse gold suspensions, *Nature (London)* 241, 20, 1973.
46. Hogg, R. J., Hodgson, A. J., Henderson, D. W., Williams, K. A., Jureidini, K. F., and Zola, H., Autigen expression during early human granulocyte development studied with immuno-electron microscopy, submitted for publication.

## FURTHER READING

### Immunofluorescence

Nairn, R. C., *Fluorescent Protein Tracing*, 4th ed., Longman, New York, 1976.

### Flow Cytometry

Herzenberg, L. A. and Herzenberg, L. A., Analysis and separation using the fluorescence-activated cell sorter (FACS), in *Handbook of Experimental Immunology*, Vol. 2, 3rd ed., Weir, D. M., Ed., Blackwell Scientific, Oxford, 1978. Note: The 4th ed., currently in preparation, will have more detailed information on flow cytometry.

Loken, M. R. and Stall, A. M., Flow cytometry as an analytical and preparative tool in immunology, *J. Immunol. Methods*, 50, 85, 1982.

## Immunohistochemistry

Bullock, G. R. and Petrusz, P., *Techniques in Immunocytochemistry,* Vol. 1 and 2, Academic Press, London, 1982 and 1983.
Pearse, A. G. E., *Histochemistry, Theoretical and Applied,* Vol. 1 and 2, 4th ed., Churchill Livingstone, Edinburgh, 1980.
Polak, J. M. and Varndell, I. M., *Immunolabelling for Electron Microscopy,* Elsevier, Amsterdam, 1984.

## Cell Separation

Mishell, B. B. and Shiigi, S. M., *Selected Methods in Cellular Immunology,* Section II, W. H. Freeman, San Francisco, 1980.

## Preparation of Reagents for Immunostaining

Goding, J. W., *Monoclonal Antibodies: Principles and Practice,* Academic Press, London, 1983.
Mishell, B. B. and Shiigi, S. M., *Selected Methods in Cellular Immunology,* Section III, W. H. Freeman, San Francisco, 1980.

Chapter 6

# USING MONOCLONAL ANTIBODIES: SOLUBLE ANTIGENS

## I. INTRODUCTION

The introduction of the immunoassay revolutionized the determination of biologically active substances such as hormones, drugs, and enzymes. One example, which made an indelible impression in the author's mind, was the determination of insulin by radioimmunoassay. Previously, insulin was measured by a bioassay which required the induction of a diabetic coma in rats. Several animals were needed at each of several concentrations of hormone in order to achieve acceptable precision. In contrast, the radioimmunoassay is precise, rapid, and does not require the use of laboratory animals once the antisera are available. Another example, more widely appreciated, is the immunological determination of the pregnancy-associated hormone HCG. This forms the basis of rapid, reliable, and easy test kits, some of which can be used at home. Before the development of immunoassays, pregnancy testing depended on the induction of ovulation in laboratory animals.

Immunoassay achieved major increases in speed and specificity, and reductions in cost and in the use of laboratory animals. Monoclonal antibodies can improve the specificity of immunoassays and broaden their scope. However, the use of monoclonal antibodies in the assay of biologically active soluble substances has developed slowly in comparison to the uses of monoclonal antibodies as reagents for working with cells. One reason for this is that the benefits of monoclonal antibodies are rather less compelling when there are already well-established assays.

Quite apart from their use in immunoassays, monoclonal antibodies have an important role as research tools for the biologist studying soluble biological molecules. Monoclonal antibodies are markers for particular molecular structures and can be used to determine where and when these epitopes are found (tissue distribution, ontogeny, differentiation). Genetic polymorphism, evolutionary changes, and mutations of the antigenic determinants can be studied with the antibody reagents, which can also be used to detect expression of the antigen in interspecies hybrids or in transfected cells, thus, enabling the geneticist to identify the chromosome coding for a particular antigen. In the same way expression of "engineered" genes (DNA inserted into microorganisms using vectors) can be determined if a monoclonal antibody against the gene product is available. Monoclonal antibodies can also be used to purify a soluble substance from a complex mixture. These studies can make major contributions to our understanding of the functions of complex molecules, frequently the ultimate aim of biological studies.

This chapter describes some of the techniques used for the study of biological substances in solution in which monoclonal antibodies may be used to advantage.

## II. IMMUNOASSAY: DETERMINATION OF THE CONCENTRATION OF AN ANTIGEN IN A MIXTURE

### A. General Discussion

The development of the radioimmunoassay established immunoassay as a sensitive, precise, and specific technique for the determination of biologically active substances. In radioimmunoassay, a small amount of the substance to be assayed must be available pure. This pure material is radiolabeled and mixed with the sample, which contains the

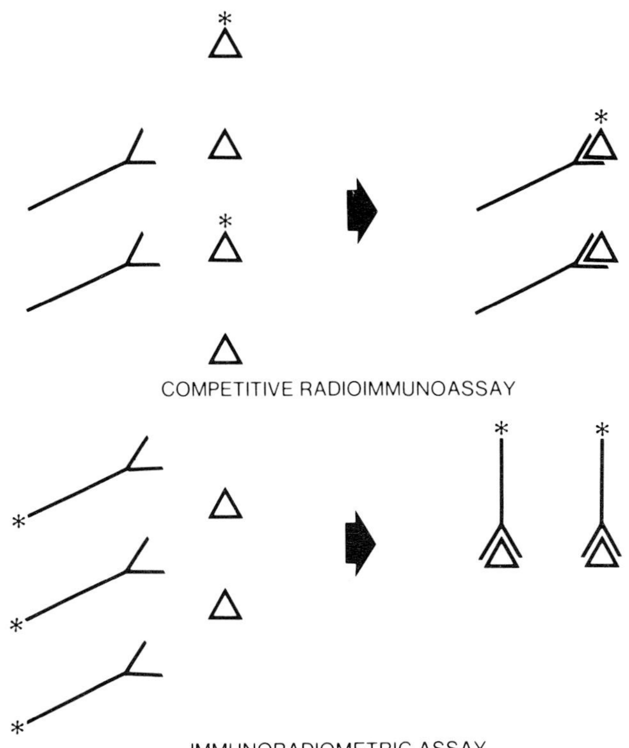


FIGURE 1. Principles of competitive radioimmunoassay and immunoradiometric assay. In the competitive radioimmunoassay pure antigen is available radiolabeled; the antigen in the sample competes with the radiolabeled antigen for the limited number of antibody molecules. In the immunoradiometric assay the antibody is labeled and added in excess; the amount of radioactivity bound is a measure of the amount of antigen. See text for detailed discussion.

same substance in unlabeled, impure form. Specific antibody is added, and the complexes formed between antigen and antibody are precipitated. The radioactivity in the precipitate is counted. Provided the amount of antibody is limiting, the substance to be determined will not all be precipitated, and the radioactive and unlabeled material will compete for the limited number of antigen-binding sites of the antibody. The higher the concentration in the sample the less radioactivity will be precipitated. The assay depends on competition for binding sites (Figure 1). The principal disadvantage of competitive radioimmunoassays is the need for highly purified material to be radiolabeled. The need to precipitate the complex from solution provides the major technical difficulty. The major strength of the method is that irrelevant antibodies present in the antiserum, which react with other components of the sample, will not interfere with the test, because they do not bind the labeled antigen.

Related assays were developed, which avoided the use of purified material and used, instead, labeled antibody to measure antigen (Figure 1). This second type of assay, also often called radioimmunoassay, but better called immunoradiometric assay (IRMA), is subject to interference from antibodies against constituents of the mixture, other than the substance being determined. Competitive radioimmunoassay has, therefore, been preferred to IRMA as being more specific. The increased specificity of monoclonal antibody, as compared to polyclonal antisera, confers on the IRMA the desired

Table 1
PUBLISHED EXAMPLES OF IMMUNOASSAYS BASED ON MONOCLONAL ANTIBODIES

| Substance assayed | Type of assay | Ref. |
|---|---|---|
| Cardiac myosin | Competitive RIA; IRMA with 2 Mabs | 1 |
| Alpha-fetoprotein | Competitive RIA | 2 |
| Glial fibrillary acidic protein | ELISA with 1 Mab and 1 rabbit antibody | 3 |
| Serum factor VIII-related antigen | Competitive ELISA | 4 |
| Bovine antibody against rotavirus | Isotype-specific ELISA using Mab against bovine Ig isotypes | 5 |
| Rheumatoid factors | ELISA with IgG-specific Mab | 6 |
| Amniotic fluid alkaline phosphatase (in prenatal diagnosis of cystic fibrosis) | ELISA — the antigen being the enzyme | 7 |

specificity. The IRMA is generally carried out as a solid-phase procedure, with the antigen immobilized on a solid carrier. This is most conveniently achieved using another antibody against the antigen to bind the antigen to the carrier. Thus, two antibodies are needed; the first is coupled to the carrier and the second is the radiolabeled detection reagent ("readout antibody").

Enzyme-linked immunoassay (ELISA) can be carried out as a competitive assay, analogous to radioimmunoassay, or as an analog to IRMA. In ELISA, the radioactive label is replaced by an enzyme-substrate reaction which produces a measurable color. As discussed already for isotope-based assays, the specificity of monoclonal antibodies confers on the solid-phase, noncompetitive form of the assay the necessary specificity, and solid-phase ELISA is now used extensively both in the determination of soluble antigen and in the measurement of antibody against soluble antigen.

The choice between isotope-based assays and ELISA is affected by a number of considerations. The use of isotopes is associated with a real hazard and increasingly tight legal control. On the other hand, isotopic assays probably still have a competitive edge in terms of sensitivity and have more predictable assay parameters — the final result (counts) is always directly proportional to the amount of isotope present in the final sample. In ELISA, because of the catalytic nature of the reaction, a certain amount of enzyme can give a small or a large amount of product, depending on the reaction conditions. Since the reaction product can inhibit the enzyme, ELISAs have a relatively narrow "dynamic range" — the graph of measured product against antigen concentration flattens out. Perhaps the most important consideration in favor of ELISA is the long shelf-life of the reagents, as compared with the short life of iodinated material (the radioactive half-life of $^{125}I$ is 60 days). This means that only assays which are carried out frequently can sensibly be set up as isotopic assays.

There is a large number of immunoassays based on polyclonal antisera available commercially and working well in routine laboratories. The only reason to replace such assays with monoclonal-antibody-based assays is the availability of an unlimited source of reproducible reagent. Against this, not all monoclonal antibodies work well in immunoassay. Some have low affinity, and the presence, in polyclonal preparations, of several antibodies against different determinants of the antigen makes for a stronger reaction.

Table 1 lists a number of published examples of immunoassays using monoclonal antibodies. The ELISA is discussed in detail in the following section, with particular reference to its use for determination of antigen. The use of the ELISA to screen for monoclonal antibodies is discussed in Chapter 3 (Section VIII.B), but the methodology is similar.

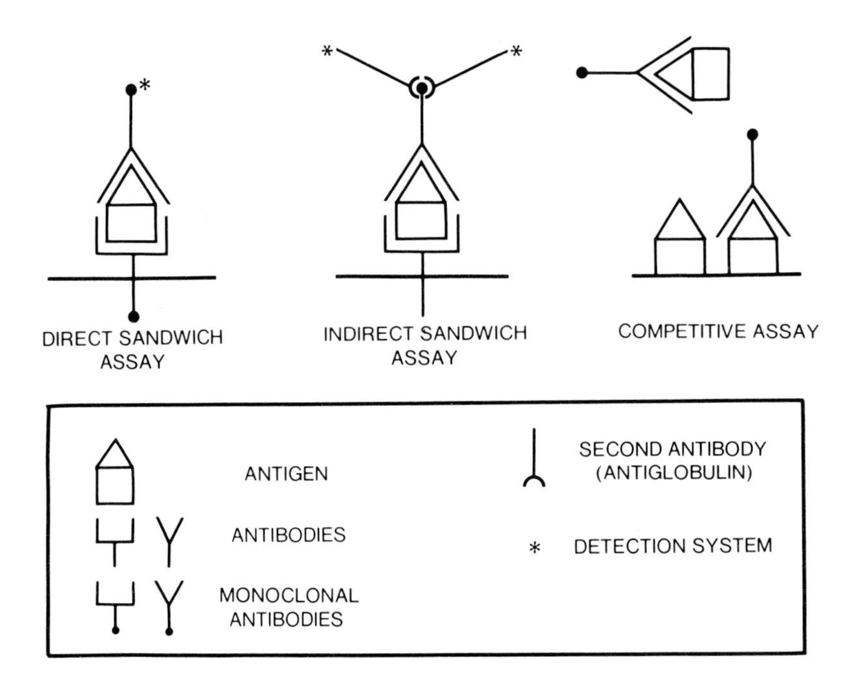


FIGURE 2. Three different solid-phase immunoassay configurations. In the sandwich assays the antigen is bound by an antibody-coated surface. A second antibody against the same antigen then binds to the antigen, allowing quantitation (with radioactive tag or enzyme reaction). The second antibody may be directly labeled (direct sandwich assays) or may itself be measured using a labeled antiimmunoglobulin (indirect sandwich assay). In the competitive assay free antigen in solution competes for antibody sites, reducing the amount of antibody binding to the solid phase. The amount of antibody bound to the solid phase provides a measure of the amount of antigen in solution.

## B. Enzyme-Linked Immunoassay (ELISA) — Technical Aspects

### *1. Introduction*

The principle of the ELISA as used to determine either antigen or antibody is illustrated in Figure 2. For the noncompetitive ELISA for antigen, two antibodies directed at distinct epitopes on the antigen are used: one to bind the antigen to the solid matrix and the other to detect the antigen and measure its concentration. As indicated in Figure 2, the ELISA can take several different forms. If two monoclonal antibodies against different epitopes on the antigen are available the direct sandwich assay (an ELISA analog of the IRMA) may be used. The indirect sandwich assay is capable of greater sensitivity, because more enzyme molecules are bound per epitope. Furthermore, the indirect assay can make use of commercially available enzyme conjugates of antimouse immunoglobulin, and so avoids the need to couple the monoclonal antibody to enzyme (see Chapter 5, Section III.C.2 on the problems inherent in enzyme-antibody coupling). However, the indirect assay cannot readily be carried out with two mouse monoclonal antibodies, since the antiglobulin reagent would not be able to distinguish between the two monoclonal antibodies and would bind directly to the first layer, irrespective of the presence or absence of antigen. The indirect ELISA can be used either by having monoclonal antibodies which can, in some way, be distinguished from each other, or by using an antibody from a different species as one of the antibody layers. For example, the second monoclonal antibody can be biotinylated, in which case an avidin system is used to detect the binding of the biotinylated antibody. These systems are analogous to the variations in immunoenzyme detection systems, discussed in detail in Chapter 5, Section III.C.2. If the ELISA is based on one monoclonal antibody and a polyclonal antibody preparation from a different species, specificity need not be sacrificed. Although the polyclonal antibody may bind components of the mixture, other than the substance being measured, the monoclonal antibody step should ensure adequate specificity. The assay for glial fibrillary acidic protein referred to in Table 1 is an example of this type of ELISA.

As indicated in Figure 2, the ELISA can also take form of a competitive assay. An example of this ELISA is the assay for factor VIII-related protein (Table 1).

There are many different technical variations possible in ELISA. The following section provides technical details as used in the author's laboratory, followed by notes on some of the variables.

*2. Materials*

*a. ELISA Plates*

These are 96-well flat-bottomed microtiter plates, marketed as special ELISA-grade plates. In some assays the cheaper standard plates will work, but generally they will not, and even the ELISA-grade plates have to be checked for suitability for the particular assay. The desirable properties are

1. Adequate optical quality if the plates are to be read quantitatively. Determine by filling wells with a colored solution and reading the plates in an ELISA plate reader, checking for variations in optical density from well to well and plate to plate. This property may vary from batch to batch from the same manufacturer, since it depends on the age and condition of the manufacturing equipment. Generally, however, ELISA-grade plates from major manufacturers are satisfactory in this regard.
2. Binding capacity will depend on the ligand being bound: antibody, protein antigen, lipid antigen, carbohydrate, virus, etc. Most ELISA work is carried out with protein antigen (to detect antibody) or with antibody bound first (to measure antigen). When setting up a new assay it is advisable to compare plates from different manufacturers and chose the one most suited to the ELISA being developed. The range of concentrations of the test material which can be measured, and the slope of the assay, should be determined.

In routine use it is often desirable to run only a few samples, and some manufacturers sell ELISA wells in rows which are separate, but fit a plate-sized carrier. It is important to check that these will fit properly into the ELISA plate reader.

*b. ELISA Plate Reader*

These instruments are essentially photometers designed to read each well in a 96-well plate, automatically moving on to the next well, reading the whole plate or those rows and columns used. Several commercial instruments are available. Compatibility with a small computer, and availability of a suitable program to process the results and transform optical density readings into concentrations of ligand are important additional items to look for when selecting an instrument.

*c. Buffers*

1. Coating buffer: carbonate/bicarbonate buffer, pH 9.6, made up as follows: dissolve 1.59 g $Na_2CO_3$ in 400 mℓ distilled water, and 2.93 g $NaHCO_3$ in a further 400 mℓ distilled water and mix. Add sodium azide ($NaN_3$; 0.2 g) and check pH. If necessary, adjust pH to 9.6 by addition of carbonate or bicarbonate solution;

make up to 1 ℓ. Store at 4°C. Although it has become common practice to coat proteins, including antibody, in high-pH buffer, it is often possible to coat at physiological pH. If there is any reason to suspect that the material being coated is unstable at high pH, it is worthwhile experimenting with neutral pH.

2. Blocking/washing solution — PBS/tween 20. Weigh out the following: NaCl, 160 g; $KH_2PO_4$, 4 g; $Na_2HPO_4 \cdot 12H_2O$, 58 g (or anhydrous, 23 g); KCl, 4 g; $NaN_3$, 4 g. Dissolve each in 250 mℓ, mix, add 10 mℓ tween 20 and make up to 2 ℓ. This is a 10× stock solution and should be stored at 4°C.

*d. Antibodies*

As discussed above, the ELISA for antigen determination commonly makes use of two antibodies reacting with different epitopes. In order to ensure specificity, at least one of these antibody preparations should be monoclonal. If both are monoclonal and from the same species, one of them must either be coupled directly to enzyme or derivatized in such a way that it can be picked up by an appropriate enzyme-labeled reagent (for example, avidin-enzyme).

The first antibody, to be coated onto the plate, should be relatively pure, so that the available protein-binding sites on the plastic are not occupied by irrelevant protein. Thus, if the first layer is to be polyclonal antibody it should be affinity purified. It is usually much easier to use the purified immunoglobulin from the monoclonal antibody ascites (there is no need to affinity purify the monoclonal) as the first layer, and the polyclonal antibody as the second antibody. In this case it is not even necessary to prepare the globulin fraction from the polyclonal antiserum.

*e. Detection Reagents*

Conjugated antiglobulin reagent — Whether the ELISA is being used to determine antigen using a monoclonal antibody or to detect monoclonal antibody in a screening assay, the next layer consists of an enzyme-conjugated antibody directed against immunoglobulin. If the assay is to screen hybridomas, the antiglobulin clearly must be directed against mouse immunoglobulin, while in an antigen assay, the antiglobulin must be directed against whatever species is used for the second antibody (see Figure 2). The enzymes most commonly used are horseradish peroxidase and alkaline phosphatase. Preparation of conjugates is fairly straightforward, but they can be obtained as good-quality reagents from various suppliers of immunochemicals. These reagents have been discussed in detail in Chapter 5, Section III.C.2, in the context of immunoenzyme histological techniques. At least in the early stages of setting up a new assay, it would be advisable to obtain a reagent marketed specifically for ELISA.

Substrate — These are available commercially. For peroxidase the substrate most commonly used is *o*-phenylenediamine, which should be handled as a carcinogen. *p*-Nitrophenyl phosphate is the substrate used most often with alkaline phosphatase-linked antibodies. This substrate is available in convenient tablet form (Sigma 104 phosphatase substrate tablets; *p*-nitrophenyl phosphate disodium salt, 5 mg/tablet). For use, make up substrate buffer as follows: $NaN_3$, 0.2 g; $MgCl_2 \cdot 6H_2O$, 100 mg; dissolved in 800 mℓ distilled water. Add diethanolamine, 97 mℓ, slowly and with mixing. Adjust pH to 9.8 with 1 *N* HCl. Make up to 1 ℓ with distilled water. Store at 4°C. Substrate is dissolved in the buffer to give a concentration of 1 mg/ℓ.

*3. Method*

The procedure described below is suitable for the determination of antigen; the changes needed to use the ELISA to measure antibody are discussed in Note d following the "Method" description.

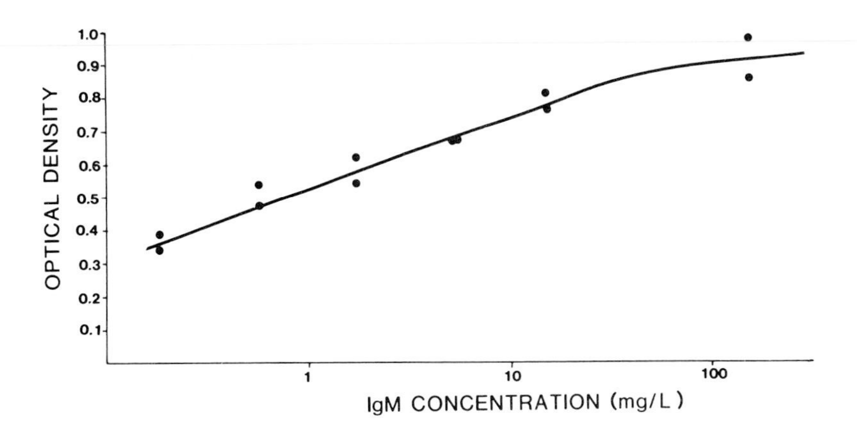


FIGURE 3. ELISA calibration curve. ELISA for human IgM: monoclonal antibody against human IgM was purified and applied to the plastic ELISA plate. The plate was washed and blocked. Human IgM-containing samples (culture supernatants) were applied, the wells were washed, and human immunoglobulin was measured using a rabbit-antihuman immunoglobulin conjugated to alkaline phosphatase. See text for details. The calibration curve was obtained using a sample of known IgM content.

1. Antibody: dilute in coating buffer to a concentration of 10 to 100 $\mu$g/m$\ell$ (for a purified monoclonal antibody, but the best concentration will have to be determined empirically for the assay in question). Add 100 $\mu\ell$ to each well of the ELISA plate and leave overnight in a humid chamber at 4°C. Remove the antibody solution (which may be reused) and wash the plate twice with blocking/washing buffer. Washing may be done rapidly by "flicking" out solutions and flooding wells from a pipette. ELISA plate washers, which aspirate and wash one row at a time, are a convenient aid (for example, the Immuno-Wash 12, Nunc InterMed, Roskilde, Denmark). Each wash should be allowed to stay in contact with the plate for 2 min. The antibody-coated plate may be stored frozen after flicking out the buffer and wrapping the plate in plastic film.
2. Test sample: if using a stored plate, allow it to warm up and wash once as above. Add sample, 100 $\mu\ell$/well. The concentration will have to be determined empirically (see Note a). Include suitable negative controls and positives of known concentration to plot a standard curve. Incubate 2 hr at 4°C (see Note b) and wash five times in wash buffer, as described above.
3. Add second antibody — usually rabbit antibody against the antigen. The concentration must be determined empirically, in a checkerboard titration in which the antigen concentration is also varied; see Note a. Incubate 2 hr at 4°C; wash five times.
4. Add conjugated antiglobulin: incubate 2 hr at 4°C and wash five times. The concentration will have to be determined empirically, using the supplier's data sheet as a guide if a commercial reagent is used.
5. Add substrate, incubate 30 min at 37°C, and read in ELISA plate reader.

An example of a calibration curve for an ELISA for human IgM is shown in Figure 3.

### *4. Notes*

#### *a. Checkerboard Titrations to Establish Optimal Concentrations*

The optimal concentrations of the various reactants will vary greatly, depending on the reactants, the plastic tray used, and the conditions of the assay. Concentrations given above provide a starting level, but it is important to check the optimal concentration range for each reactant by carrying out "checkerboard" titrations — where two reagent concentrations are varied, one along the rows and the other down the columns.

*b. Incubation Times*

The kinetics of antigen-antibody interaction are similar in principle to any other association-dissociation reaction, and the state of equilibrium is governed by the law of mass action. At equilibrium, lower temperatures will favor binding, while higher temperatures will favor dissociation. However, equilibrium is reached more rapidly at higher temperatures. With any particular antigen-antibody pair, it is impossible, without knowing the rate constants for association and dissociation, to predict how long it will take to reach equilibrium. It may, thus, be advantageous to experiment with incubations at 20 or 37°C if the conditions given above do not work well. Alternatively, incubation periods at 4°C may be increased. A risk in using higher temperatures is that the first layer will dissociate from the solid matrix.

*c. Alternative Detection Systems*

If the sensitivity of the labeled antiglobulin method described here is inadequate for a particular application, it may be worth trying to increase sensitivity (at the cost of increasing the complexity and length of the assay) by adapting either the enzyme-antienzyme or the biotin-avidin systems described for immunohistochemistry in Chapter 5 (Section III.C). Alternatively, the use of iodine-labeled antiglobulin remains a simple and sensitive method. The relative merits of radioisotope and enzyme-based assays have been discussed above.

*d. Modification of ELISA for the Detection of Antibody*

As discussed in Chapter 3 (Section VIII.B.2), the ELISA is well suited to the screening of hybridoma supernatants. The general features of the assay can be the same as described above, except that the antigen is applied directly to the plate. Note that the antigen should be as pure as possible for this purpose, since the assay can tell us only that the hybridoma supernatant is reacting with material on the plate; the assay does not distinguish between the antigen of interest and an impurity.

ELISAs are suitable for the detection and measurement of antibody in pathological sera — for instance, to determine the response to immunization or to assess autoimmunity. Such assays do not involve monoclonal antibodies, but it is often desirable to determine if the patient (or animal) is responding to recent first exposure (characteristically, an IgM response) or has previously been exposed (IgG response). Monoclonal antiimmunoglobulins are useful in this situation, either in a "capture assay" to select the appropriate immunoglobulin class from the serum, or as the last antibody layer, to detect only the appropriate antibody class (Figure 4).

C. Dot Immunoblotting on Nitrocellulose

*1. Introduction*

ELISA can be carried out by absorbing the first reactant on nitrocellulose instead of plastic. This technique has come to be called "dot immunoblotting" by analogy with the dot hybridization technique for nucleic acid analysis. The dot-ELISA is particularly useful when the sample is limiting, because smaller volumes can be used. Although reflectance-spectrophotometers are available, the method is probably less precise than ELISA in plastic wells, and is most useful as a qualitative and semiquantitative test. For example, the method described below can be used for semiquantitative determination of antigen concentration, by spotting serial dilutions of antigen extract onto the nitrocellulose. Within limits, which can be determined empirically, the protein is taken

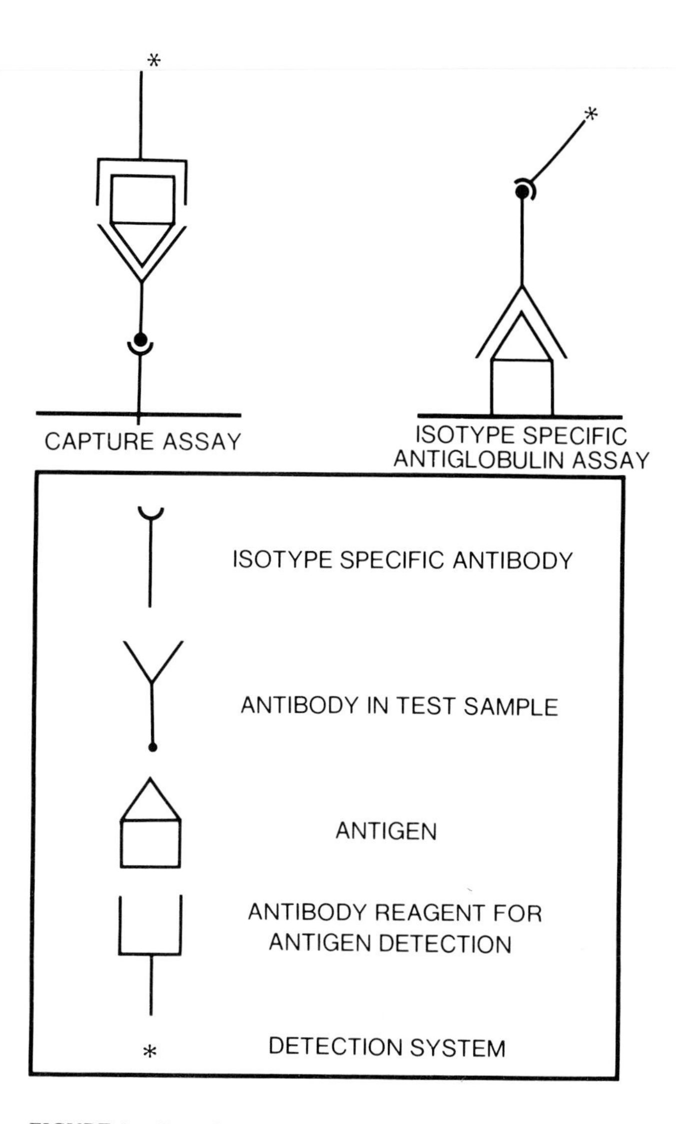


FIGURE 4. Two alternative assays to measure antibody of a particular class in patient sera. Example: determination of IgM antibody against rubella virus in serum sample, to detect recent exposure to the virus. In the capture assay, monoclonal antibody against human IgM is bound to the solid phase. This "captures" IgM from the test sample. If there is any antirubella antibody in the IgM, virus added in the next step will be bound and can be detected with a labeled antiviral antibody. In the isotype-specific antiglobulin assay the virus is bound directly to the solid phase. Antiviral antibody (of any class) from the patient serum binds to the virus. Any bound IgM is then detected using labeled monoclonal antihuman IgM. The two assays have advantages and disadvantages, depending on the intended use.

up quantitatively. When the ELISA is carried out on plastic the unknown should not be applied directly to the plastic because the proportion bound is low and variable.

Radioactive isotope may be used instead of enzyme, and the many variations in configuration of the assay, discussed above for the ELISA in plates (Section B) and in more detail for immunohistochemistry (Chapter 5, Section III.C), can also be applied to dot-immunoblotting.

A procedure used in the author's laboratory is described below, followed by notes on alternative detection systems.

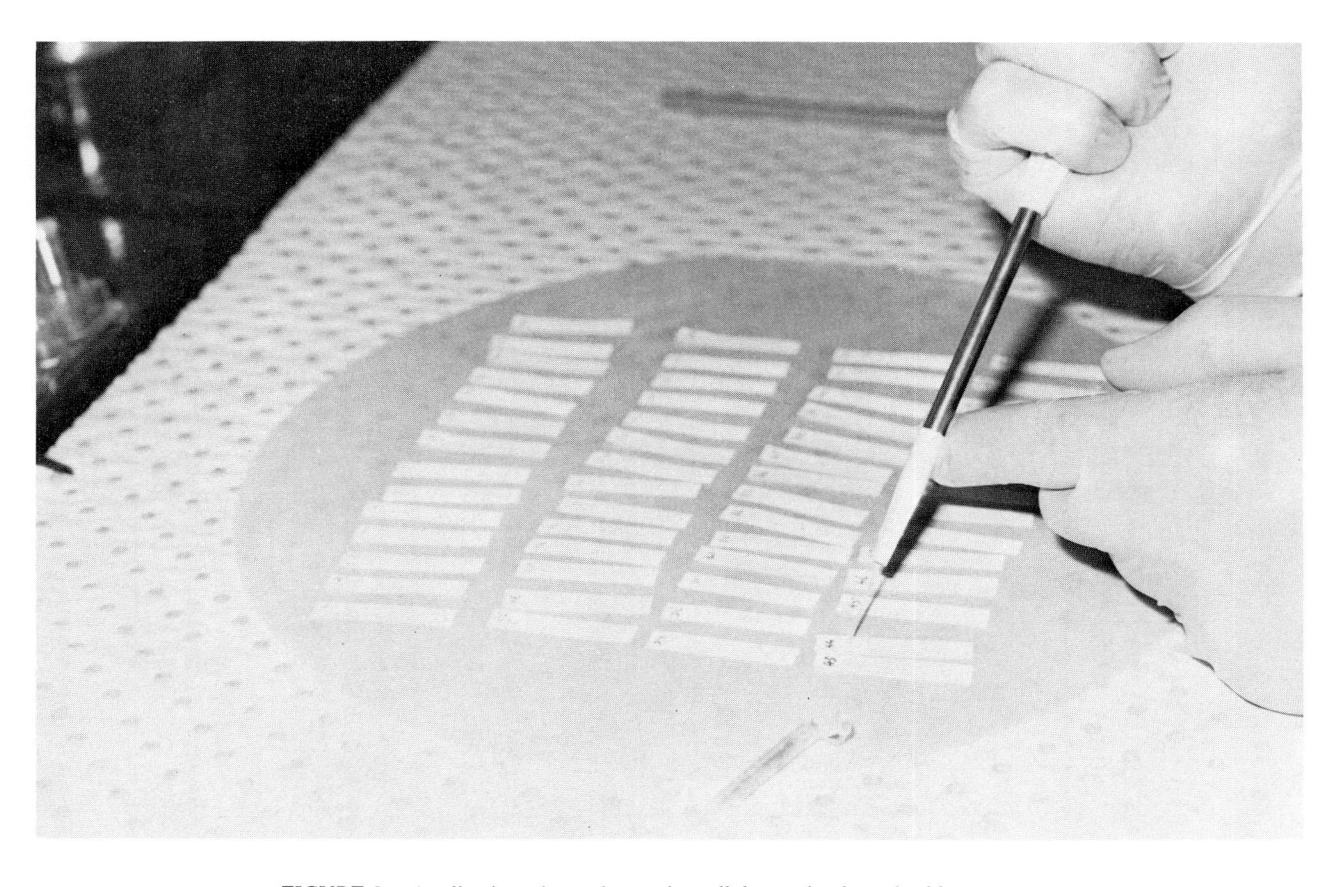

FIGURE 5. Application of samples to nitrocellulose strips for a dot-blot assay.

*2. Materials*

1. Nitrocellulose: material prepared specifically for blotting is available from a number of manufacturers; pure nitrocellulose filtration membranes (but not membranes made of composite materials) are also suitable. The nitrocellulose may be cut into whatever shape is required; one way to use it is in an apparatus designed to handle 96 samples in a configuration similar to that of a 96-well ELISA plate (for example, the "Bio-Dot" apparatus from Bio-Rad Laboratories). When handling smaller numbers of samples, a convenient arrangement is to work with small strips of nitrocellulose (0.5 × 4 cm) which can accommodate five to six spots on each strip (Figure 5). The strips can then be washed and stained in small tubes, for maximum economy of reagents. Nitrocellulose must be handled cleanly and with gloves; the amount of protein in a fingerprint will be clearly visible.
2. Tris-buffered saline: 0.05 *M* tris/0.15 *M* NaCl, pH 7.6.
3. Blocking buffer: 6% suspension of skim milk powder in PBS, centrifuged at 5000 × G for 15 min and filtered through a 0.8-μm followed by a 0.45-μm membrane.
4. Bridging antibody: goat-antimouse immunoglobulin, fractionated by octanoic acid precipitation (Chapter 5, Section V.A). This reagent must be titrated; a dilution in the range 1/20 to 1/2000 should be suitable.
5. Alkaline phosphatase-antialkaline phosphatase reagent (APAAP): this reagent is described in Chapter 5, Section III. The complex is prepared by mixing 1 vol of a solution of 50 mg/mℓ alkaline phosphatase (Sigma P3877 or equivalent) with 9 vol of antialkaline phosphatase monoclonal antibody FMC55 tissue culture supernatant. The complex is available from DP Diagnostics. Alternative detection reagents are discussed in Note a. The mixture is incubated 2 hr at room temperature or overnight at 4°C before use, to allow the complex to stabilize. The complex is stable at 4°C for at least 1 year.
6. Substrate for alkaline phosphatase: Sigma naphthol AS-BI phosphate or naphthol AS-MX phosphate; 5 mg dissolved in 0.2 mℓ dimethylformamide in a glass tube. Once the substrate has dissolved, make up to 10 mℓ with 0.1 *M* tris buffer, pH 8.2. If the test material contains endogenous alkaline phosphatase, this enzyme may be inhibited by the addition of 1 m*M* levamisole. Immediately before use add 10 mg Fast Red TR salt (Sigma), and filter the solution straight into the tubes containing the strips to be stained.

*3. Method*

1. Spot 1-μℓ aliquots of serial dilutions of antigen solution on to nitrocellulose (Figure 5) and allow to dry.
2. Block remaining protein-binding activity by immersing the nitrocellulose strip in blocking buffer for 30 min at 37°C. Wash three times in tris-buffered saline, allowing about 5 min for each wash.
3. Add the antibody. Tissue culture supernatant may be used undiluted or diluted 1:1 with tris-buffered saline. Background staining may be reduced by including 1% skim milk in the diluent. Incubate overnight at 4°C, with gentle agitation.
4. Wash three times in tris-buffered saline.
5. Add bridging antibody; leave for 45 min at room temperature.
6. Wash three times in tris-buffered saline.
7. Add APAAP complex; leave for 1 hr at room temperature.
8. Wash three times in tris-buffered saline.

9. Add substrate solution and leave until stained (usually 15 to 30 min; may be left overnight but the background will increase).

*4. Notes*

*a. Alternative Detection Systems*

Binding of monoclonal antibody can be detected by a variety of reagent systems, as is the case for antigen bound to plastic tubes or wells. For instance, $^{125}$I-labeled anti-mouse immunoglobulin or $^{125}$I-labeled protein A may be used. Antimouse immunoglobulin conjugated directly to alkaline phosphatase or to peroxidase may be used, together with appropriate chromogenic substrates. The biotin-avidin peroxidase system (for example, the Vectastain ABC system supplied by Vector Laboratories) is particularly sensitive. The substrates selected must yield an insoluble product, unlike those used for ELISA in tubes or microtiter wells. The immunogold technique (Chapter 5, Section III) may also be applied to blots. Under optimal conditions, isotopic methods and the most sensitive nonisotopic methods can detect picogram quantities of antigen. A number of suitable reagents are available commercially (for example, the range of enzyme-conjugated antibodies, substrates and substrate kits, and complete assay kits marketed specifically for blotting applications by Bio-Rad Laboratories). Alternatively, appropriate reagents may be prepared as described in Chapter 5, Sections III and V.

*b. Alternative Blocking Solutions*

Protein solutions such as bovine serum albumin (3%), fetal calf serum (1 to 5%) or gelatin (1%), or mixtures of these proteins may be used to block nonspecific binding of antibody to the nitrocellulose. The efficacy of blocking is assessed on the basis of the background staining intensity.

## III. BIOLOGICAL ACTIVITY: RELATIONSHIP OF BIOLOGICALLY ACTIVE SITE TO IMMUNOGENIC EPITOPE

When an animal is immunized with a biologically active molecule derived from a different species, the recipient animal will usually possess a similar molecule carrying out the same function. The immune response will be directed against minor differences, such as regions of amino acid sequence which have not been conserved during evolution. These regions will not necessarily be associated with the biologically active site of the antigen. Indeed, since function of many enzymes or hormones is conserved, it might be predicted that the active sites would be nonimmunogenic, at least compared with the rest of the antigen molecule.

In practice, some monoclonal antibodies inhibit the activity of the antigens they react with, many do not, and others stimulate activity. Inhibition, where it occurs, is usually incomplete, indicating that the binding epitope is not central to the active site. This topic is not discussed further here, but the reader is referred to the review by Harris[8] which discusses a number of examples.

## IV. IMMUNOCHEMICAL CHARACTERIZATION OF THE ANTIGEN REACTING WITH A MONOCLONAL ANTIBODY

### A. Introduction

In principle, we set out to make a monoclonal antibody against a particular molecule and devise a screening procedure which ensures that we select antibody of the right specificity. In practice, we often obtain monoclonal antibodies which react with an unknown component of a mixture, and we wish to identify and characterize the antigen. Alternatively, if the screening process suggests we do have antibody against the antigen of interest, we may wish to confirm this immunochemically. Exchange of information between laboratories is greatly facilitated if antibodies can be shown, immunochemically, to be directed against the same or distinct antigens.

In the first instance, we need to know what sort of molecule the antibody reacts with (protein, lipid, carbohydrate, etc.). If the antigen is a protein, we will usually wish to determine its molecular weight, whether it is glycosylated and whether it is a single-chain or multichain protein. Beyond this preliminary level of characterization, an antigen will have to be purified by affinity methods (Section V) in order to determine its composition and structure in more detail.

This section will describe key methods used in the characterization of protein and glycoprotein antigens. There are many variations described in the literature, and no single method is guaranteed to work with all antigen-antibody systems. References will be given to enable the reader to find alternative methods, if the techniques described here do not work well with the reader's antigen and antibody. Much of the published work on monoclonal antibodies deals with protein antigens, and the emphasis in this section will be on proteins and glycoproteins.

### B. Principles of the Methods Used

#### *1. Nature of the Antigen*

In many situations it is fairly safe to assume that the antigen is a protein; for instance, if the antibody was made after immunization with a pure protein or a mixture of proteins. It is, thus, often possible to go directly to techniques for molecular weight detection. A common, and acceptable, approach is to ask whether or not the antigen is a protein only if there is any doubt, for example, if repeated attempts to characterize a protein by the methods described in the following sections have failed.

The procedures used to determine the nature of the antigen are straightforward, and details are not presented here. The antigen may be subjected to lytic enzymes, physical insult (boiling) or chemical treatment, and loss of antigenicity determined by the assay used in screening for antibody activity. The results must be interpreted cautiously. Protein conformation is usually lost on boiling, but not always. Broad-specificity proteases such as pronase can degrade many proteins to amino acids, but in a native protein some parts of the molecule may not be accessible to the enzyme. Enzyme preparations are often contaminated by other enzymes — nucleases by proteases, for example. Glycosidases are very selective in the bonds they hydrolyze, and chemical procedures which damage carbohydrate epitopes, such as periodate oxidation, are also selective. Lipids may be characterized to a limited degree by selective extraction into organic solvents.

If it is possible to work with a system in which the antigen is synthesized in vitro, confirmation of the nature of an antigen may be obtained by using selective inhibitors of particular biosynthetic processes. Cycloheximide and puromycin inhibit protein synthesis; tunicamycin inhibits glycosylation of protein or lipid. Again, caution must be exercised in interpreting the results. The biosynthesis of an antigen may depend on an enzyme which has a rapid turnover; cycloheximide may inhibit the synthesis of the antigen, not because the antigen is protein, but because the synthesis of the enzyme is inhibited.

Perhaps, because of these limitations which must be placed on interpretation, studies of this kind are rarely reported. Nevertheless, they can be useful, and this type of analysis, as applied to cell membrane antigens, has been discussed briefly in Chapter 5 (Section II.D.2) and described in detail elsewhere.[9]

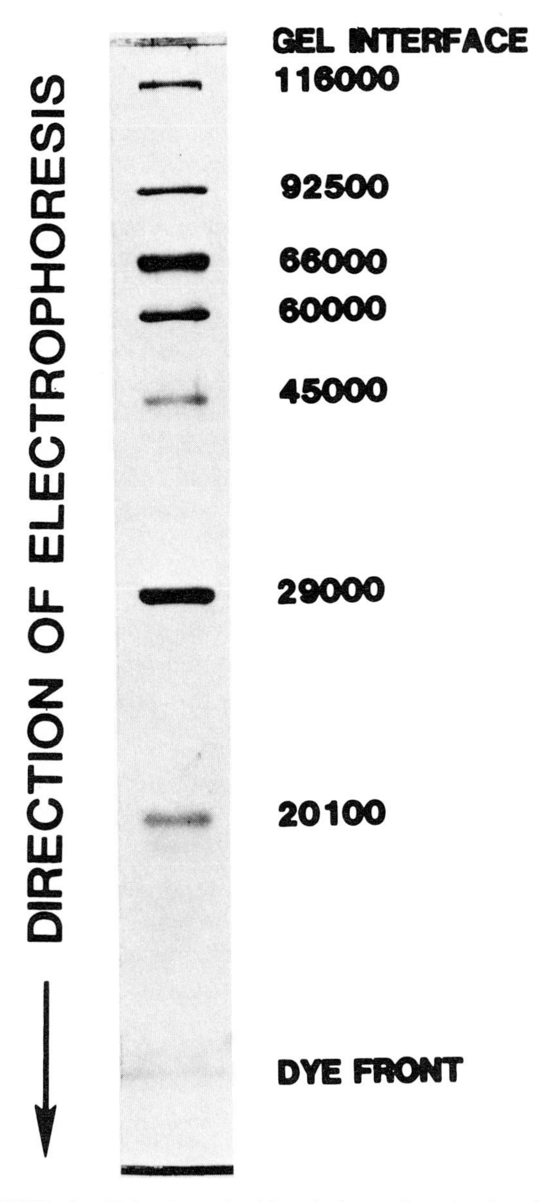


FIGURE 6. SDS-polyacrylamide gel electrophoresis of molecular weight standards. Standards (with MW in brackets) are, from top: β-galactosidase, phosphorylase B; bovine serum albumin; catalase; ovalbumin, carbonic anhydrase, and trypsin inhibitor.

*2. Protein Molecular Weight and Characterization*

*a. Electrophoresis in Gel*

Protein molecular weights are determined almost universally by electrophoresis in polyacrylamide gel. The protein is treated with the anionic detergent sodium dodecyl sulfate (SDS) to form protein-detergent micelles. The ionic detergent cancels out charge differences between proteins, and the rate of migration of the micelles in the electric field is proportional to the Stoke's radius of the micelle, which is, in turn, proportional to the molecular weight of the protein. Stoke's radius is a measure of size rather than molecular weight, but, provided the protein has been denatured by boiling, the migration of the protein is approximately related to the molecular weight (MW) by the equation:

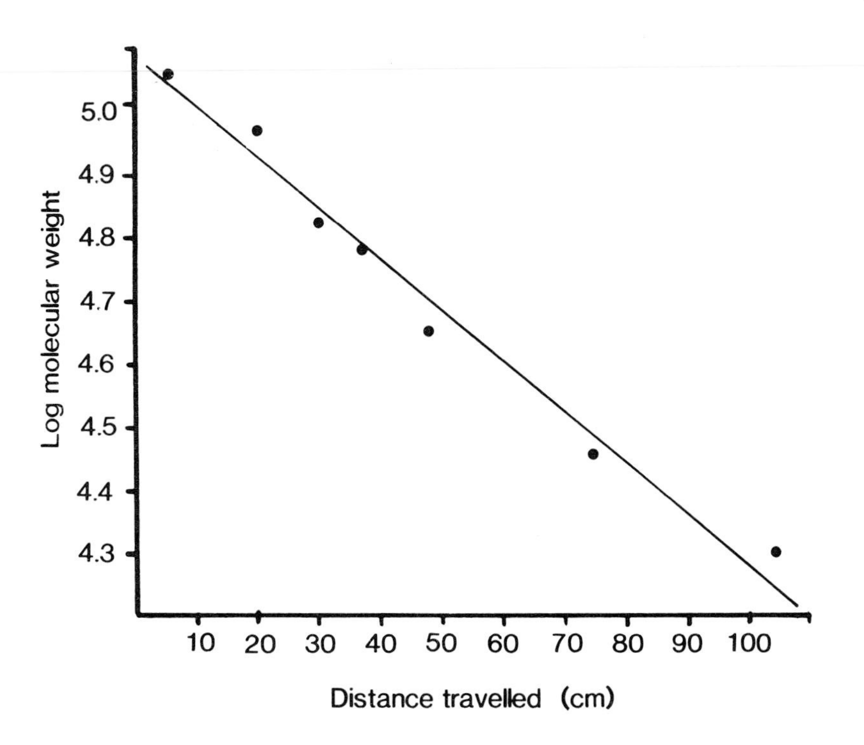


FIGURE 7. Calibration graph for determination of apparent molecular weight — plot of $\log_{10}$ MW against mobility for the gel shown in Figure 6.

$$\log(MW) = k \cdot D$$

where k is a proportionality constant and D is the distance migrated from the origin. The MW of an unknown protein may be determined by running a series of standards of known MW (Figure 6) and plotting a graph of log(MW) against D for the standards (Figure 7). The relationship is only approximate, and proteins which are highly glycosylated, highly charged, unable to bind sufficient SDS, or retain a degree of assymetry after denaturation will have anomalous mobilities. Molecular weights determined by electrophoresis should be referred to as "apparent" molecular weights. Since the molecular mass is not determined absolutely, but by comparison with standards, the term "relative molecular mass" is appropriate,[10] and units are not appropriate. However, results obtained in this way are often described as molecular weight in daltons.

When, as is usually the case, the sample being studied is a mixture of proteins, electrophoresis will yield a complex series of bands. Which of these bands is the antigen detected by the monoclonal antibody? There are two distinct approaches to answering this question. One is to detect the antigen after electrophoresis, using the antibody as a probe; the other is to use the antibody to purify the antigen before electrophoresis.

*b. Probing for the Antigen — Electrophoretic Blotting*

Although it is possible to detect antigen after electrophoresis by overlaying the gel with antibody, the method which has proved the most satisfactory to date involves making a blot — laying a sheet of nitrocellulose over the electrophoretic gel and transferring protein from the gel to the nitrocellulose. The nitrocellulose replica, or blot, is then reacted with antibody, and the site of antibody binding is identified either with a radioactive antiglobulin or by an enzyme method, as described previously for dot-immunoblotting (Section II.C). The blotting (transfer of protein) is best effected by electrophoresis, and equipment designed specifically for electrophoretic transfer is available commercially (Bio-Rad Laboratories).

Electrophoretic blotting (also called Western blotting, a play on names for cognoscenti of the techniques for blotting nucleic acids — Southern and Northern blotting) has so far proved disappointing as a technique for the identification of antigens recognized by monoclonal antibodies. The principal reason for this, almost certainly, is that blotting has been carried out after SDS electrophoresis of boiled/reduced samples, and few antigenic epitopes of proteins will survive this treatment. Epitopes which do survive include carbohydrate groups and determinants which depend entirely on the amino acid sequence, rather than on the tertiary structure, of the protein. A major advantage of the technique, over the precipitation or purification techniques discussed in the next section, is that blotting should reveal which chain of a multichain protein carries the antigenic determinant. The technique is certainly worth trying with any new antibody, and the probability of successfully identifying the antigen may be improved by attempting to "renature" the antigen after electrophoresis[11] and by using different types of membrane (see Bio-Rad catalog). It is also possible to avoid the extreme conditions of SDS electrophoresis, boiling and reduction, but due consideration must be given to the purpose of the experiment. If the protein is not boiled in SDS the mobility will not give an indication of the molecular weight. Alternative procedures may be used. Gradient-pore electrophoresis will separate proteins according to size without the need for detergent, but the "size" measured will be highly dependent on the shape of the molecule, unless it is first denatured. Isoelectric focusing and electrophoretic separation according to the intrinsic charge of the protein may be used to characterize an antigen and compare it with a known protein.

To summarize, electrophoretic blotting has not been universally successful with monoclonal antibodies, particularly against membrane proteins, but it is a new technique which probably has much unrealized potential.

### *c. Prior Purification of the Antigen — Immunoprecipitation and Immunoaffinity Purification*

Antibody provides a specific reagent enabling the purification of the antigen. If extensive studies are intended on the composition and structure of the antigen, it will be necessary to purify it using an affinity column, as described in Section V. If we need only enough to run an electrophoretic gel in order to measure the molecular weight and examine the chain composition of the antigen, it is often possible to achieve this by immune precipitation. The monoclonal antibody is added to the mixture containing the antigen. This mixture should be labeled with radioisotope in order to distinguish the antigen from antibody and other proteins added during the precipitation. The mixture of antigen and monoclonal antibody will, under conditions of equivalence or antibody excess, form immune complexes, but these will normally remain in solution. The complex may be precipitated by adding antibody against mouse immunoglobulin. Rabbit antibody is generally regarded as being particularly suited to precipitation. To be certain of bringing the complex out of solution, fixed *Staphylococcus aureus* organisms bearing protein A on their surface may be added to precipitate the rabbit antibodies, including complexes bound by them. This whole complex is then washed extensively to remove any protein from the antigen mixture which has bound nonspecifically. The final complex contains mouse immunoglobulin, rabbit immunoglobulin, and the *S. aureus* organisms, but the only radiolabeled protein in the complex should be the antigen. The complex may now be solubilized and run on electrophoresis gels; autoradiography should reveal the protein antigen.

The procedure works well for antigen-antibody systems of high affinity. If the affinity is low it is more difficult to break up nonspecific complexes without solubilizing the specific complex. Many variations on the basic precipitation technique have been suggested, but they do not alter the need for a high affinity of binding. If precipitation does not work, it is probably better to make a small affinity column (see Section V) rather than expend a great deal of effort fine-tuning the precipitation procedure. One advantage of the column procedure is that it may be possible to avoid radiolabeling the antigen. In a column procedure the antibody is covalently bound to the column matrix, and the only protein eluted should be the antigen. If a sufficiently sensitive method is used to stain the electrophoretic gel, the antigen may be identified directly. Both immunoprecipitation and affinity column experiments are described below.

*3. Lipid Antigens*

Although techniques for the analysis of mixtures of lipids are highly developed (see for example Lee et al.[12]), the immunochemistry of lipid antigens is not as well worked out as that of the protein antigens. Lipids are not generally soluble in aqueous solutions, whereas antibodies are not soluble in lipophilic solvents. In nature, the interaction between antibody and lipid antigen occurs at the cell surface, and it is difficult to reproduce these conditions once the antigen has been solubilized. Lipid antigens may be fractionated by thin-layer chromatography and antibody overlayed onto the chromatogram (see, for example, Hansson et al.[13]); alternatively, a blot may be taken from the chromatogram (Towbin[14]). After antibody binding the mouse immunoglobulin may be located by the staining procedures described for protein blots.

*4. Carbohydrate Antigens*

Many of the monoclonal antibodies reacting with glycoproteins and glycolipids are directed against carbohydrate epitopes. The same carbohydrate determinant can occur on glycoprotein and on glycolipid in the same cell. Carbohydrate may be radiolabeled biosynthetically by cells in the presence of labeled sugars, or chemically by procedures such as the periodate oxidation/tritiated borohydride reaction of Gahmberg and Andersson.[15] The extensive studies of Hakamori's laboratory on glycolipid and glycoprotein carbohydrate determinants have utilized monoclonal antibodies and have been reviewed by Hakamori.[16]

## C. Methods

### *1. Radiolabeling*

#### *a. Iodination — Chloramine T*

##### i. Principle

Chloramine T is a mild oxidizing agent, and at slightly alkaline pH (7.5 to 8.0) it oxidizes NaI to form $I^+$ ions. These react with tyrosine residues of the protein. At the end of the reaction sodium metabisulfite, a reducing agent, is added to neutralize the chloramine T and stop the reaction, and "cold" NaI or KI is added to compete with "hot" iodide ion and reduce nonspecific uptake of radioactivity.

##### ii. Materials

- Protein to be iodinated: 1 to 2 mg, at 10 mg/mℓ in buffer (PBS or equivalent) at pH 7.5.
- $Na^{125}I$: carrier free, 0.5 to 1 mCi/mg protein.
- Chloramine T: 2 mg/mℓ in PBS; make up fresh.
- Sodium metabisulphite ($Na_2S_2O_5$): 2 mg/mℓ in PBS; make up fresh.
- NaI: 10 mg/mℓ in PBS; make up fresh.
- Micropipettes for 5, 10, 100 μℓ.
- Stopwatch.
- Disposable desalting column (Sephadex G25, Pharmacia, or Bio-Gel P10, Bio-

Rad), preequilibrated by passing through in sequence: 10 mℓ PBS, 1 mℓ FCS, and 20 mℓ PBS.
- Ice.
- Trichloracetic acid, 10% in water.
- Gamma-counter.

iii. Method

WARNING — Iodination should be carried out in a fume cupboard by experienced personnel or under supervision. In many countries the use of a special laboratory is specified by law. Two hazards are associated with the use of radioactive iodine. The first is direct irradiation with the penetrating gamma rays emitted by the isotope. The dose of gamma radiation is reduced by using shielding and working as rapidly as possible, particularly reducing the time of physical proximity to the isotope. The second, more serious hazard is that a proportion of the isotope will get into the atmosphere. If ingested, it will localize in the thyroid, producing high local radiation doses. The use of a fume cupboard is essential.

- Add the radioactive iodide to the protein solution (typically 1 mCi to 200 μℓ).
- Add chloramine T (1 mg/5 mg protein) with a mixing action and start the stopwatch.
- After 1 min add the sodium metabisulphite (1 mg/mg chloramine T) followed immediately by the NaI (1 mg/mg protein). Place the mixture in ice to lower the temperature and apply to the desalting column as quickly as possible (within 5 min).
- Elute the first peak of radioactivity (protein-associated) with PBS. It is possible to follow the progress of the radioactivity down the column with a portable crystal gamma-ray detector, or to collect volumes previously shown to contain the protein peak. A 10-mℓ PD10 desalting column (Pharmacia) will have a void volume of about 1 mℓ and the protein will come through in the next 2 mℓ. Beyond this free iodide will begin to come through. The free iodide is best disposed of in the column.
- Place the eluted labeled protein in a lead container on ice and tidy up the iodination area. Waste must be disposed of according to local rules and the area must be checked for contamination.
- Determine the fraction of the radioactivity that is protein associated. This is done by precipitation with trichloracetic acid (TCA). There are a variety of procedures; we use the following: spot 2 μℓ of the labeled protein solution on to each of two filter-paper circles (1 to 2 cm diameter or small pieces cut from a paper filter). Allow to dry (in the fume cupboard) and keep one to count total counts. Place the other one in a large glass test-tube (a boiling tube). Add 5 mℓ cold 10% TCA and leave for 10 min in an ice-water bath. Transfer the disc to 5% TCA on a boiling-water bath, and leave for 15 min. Wash twice in 5% TCA at room temperature, once in 50% ethanol and once in ether (5 min each wash). Dry the paper disc and count both the TCA-treated and untreated discs in a gamma-counter. Calculate % TCA-precipitable counts.

iv. Notes

The radioactive iodine must be less than 4 weeks old (postreference date) for good iodination. The radioactive half-life of $^{125}I$ is 60 days, but decay products appear to interfere with the iodination reaction, making the effective life shorter.

The units used here, mCi, are still widely used though officially obsolete. A Ci is the amount of isotope producing $3.7 \times 10^{10}$ disintegrations per second; the new unit is a Bq

(Becquerel), which is the amount of isotope producing 1 disintegration/sec. A recipe requiring 1 mCi can be carried out using 37 *M*Bq.

The procedure as described works well with immunoglobulin. Concentrations of chloramine T and metabisulfite may have to be adjusted if the protein is damaged by the method described. Alternatively, the iodogen method may be used (see next section). It is probably less effective, but more gentle. Some proteins do not label well by these procedures, which label tyrosine residues. A number of procedures which attach label to lysine residues have been described; see, for instance, Bolton.[17]

Chloramine T is an oxidizing agent which loses activity rapidly. If the solid is not dry and white use a new bottle, and avoid using the material at the top of the bottle, in contact with air. Make up fresh.

Prepare all solutions and ensure everything is within easy reach; if necessary rehearse the reaction sequence before starting. The length of the reaction is only 1 min, and excess time either in chloramine T or metabisulfite can damage the protein.

### *b. Iodination — Iodo-gen*

Iodo-gen is, like chloramine T, a mild oxidizing agent and the mechanism of iodination is similar. However, the reaction occurs at the wall of the tube, because the Iodo-gen is insoluble in water and coated on to the tube before iodination. The reaction is stopped simply by pouring off the protein solution, and there is no need to add reducing agent. This method is preferred to the chloramine T procedure for labile proteins. It is also well suited to the iodination of cell surface antigens *in situ*, and the procedure described here is the one we use for cell surface labeling. It may be adapted to soluble antigens by using the volume and concentration of protein, the amount of radioactivity, and the desalting procedure described for chloramine-T iodination in the previous section.

#### i. Materials

- Iodo-gen (1,3,4,6-tetrachloro-3$\alpha$,6$\alpha$-diphenylglycouril) (Pierce Chemical Company, Rockford, Ill.)
- Iodo-gen-coated tubes: dissolve Iodo-gen in chloroform at 100 $\mu$g/m$\ell$. Chloroform is toxic and must not be used outside the fume cupboard. Aliquot 1-m$\ell$ amounts into polypropylene conical tubes (Beckman Microfuge tubes — capacity of 1.5 m$\ell$), and evaporate the chloroform. This may be done by bubbling nitrogen through the solution or by leaving the tubes in the fume cupboard overnight. Iodo-gen coated tubes may be kept refrigerated for several months without noticeable loss of activity.
- $Na^{125}I$: carrier free, 200 to 500 $\mu$Ci per iodination
- Cells to be iodinated: $2.5 \times 10^7$ in 1 m$\ell$ PBS
- Lysis buffer: 1% NP40 in 10 m*M* tris, pH 7.4; 1 m*M* EDTA; 0.15 *M* NaCl; 1 m*M* phenylmethylsulfonyl fluoride (PMSF); 1% bovine serum albumin, made up as follows: 100 m$\ell$ NaCl 0.9%; add 1 m$\ell$ stock EDTA 0.1 *M* (37.224 g/$\ell$); add 1 m$\ell$ NP40 detergent (Nonidet p40 BDH, Poole, Dorset, U.K.); add 1 m$\ell$ 0.1 *M* tris stock solution (121.1 g/$\ell$); adjust pH to 7.4; store refrigerated; make up stock PMSF 0.2 *M* in propanol. WARNING: PMSF IS HIGHLY TOXIC.

When lysis buffer is required, take out the amount required and add stock PMSF 5 $\mu\ell$/m$\ell$ lysis buffer and solid bovine serum albumin to 1% w/v.

When the lysis buffer is used to make extracts for blotting or affinity column work use as above; when used for immune precipitation add bovine serum albumin, 1 mg/m$\ell$.

ii. Method

WARNING — Refer to the notes on handling of radioactive iodine in the section describing chloramine T iodination.

- Wash cells four times in PBS to remove extraneous protein; resuspend in PBS at $2.5 \times 10^7$/mℓ.
- Gently rinse out the Iodo-gen-coated tube with 1 mℓ PBS, to remove any iodogen which has flaked off the tube wall.
- Add 1 mℓ cell suspension to the Iodo-gen-coated tube.
- Add 200 to 500 μCi $Na^{125}I$, mix the cells with a Pasteur pipette, and start stop-watch.
- Leave at room temperature (lead shielded) for 20 min, mixing the cells with a Pasteur pipette every 5 min.
- Wash the cells three times in PBS, 10 mℓ each time, being careful to remove all the liquid at each wash.
- Take 5-μℓ samples of each wash supernatant for gamma counting.
- Resuspend the cells in 1 mℓ cold lysis buffer. From this point on keep everything cold to reduce proteolysis.
- Take 1-μℓ sample of suspension for gamma counting.
- Leave cells in lysis buffer 1 hr on ice. Centrifuge at low speed to remove debris, then ultracentrifuge 1 hr at 100,000 × G to remove all particulate matter. Collect supernatant, determine counts and %TCA-precipitable counts (see previous method). Use the supernatant for precipitation immediately or store at −80°C.

iii. Notes

The volume of Iodo-gen solution used to coat the tube should be the same as the volume of material to be iodinated, since the reaction takes place at the surface of the tube.

Count aliquots of all washes to ensure that the washing is adequate. Some iodine labels lipid and enters the cell, so that, in spite of washing, only about 20% of the counts in the final supernatant are TCA precipitable.

Counts from a typical experiment, using 500 μCi $^{125}I$, are set out below as a guide: washings — total cpm in first wash, $62 \times 10^7$; second wash, $28 \times 10^7$; third wash, $7 \times 10^7$.

Lysate (before ultracentrifugation): $5.8 \times 10^7$ cpm total, 19% TCA-precipitable.

Lysate (after ultracentrifugation): $4.5 \times 10^6$ cpm, 17% TCA-precipitable.

### c. $^{35}S$ Methionine Incorporation

i. Materials

- $^{35}S$-methionine of high specific activity: SJ204 from Amersham, U.K.
- Methionine-free culture medium: make up from published formulas or use "Select-amine" (GIBCO) — media constituents with separate amino acids, designed to make up media deficient in individual amino acids.
- Dialyzed serum: to avoid swamping the radioactive methionine with free methionine from serum, the cells are cultured either in serum-free medium or in medium with serum which has been dialyzed against PBS (two changes of 20 volumes, 24 hr each) to remove the methionine.

ii. Method

WARNING — $^{35}S$ is a β-emitter. The hazard from direct radiation is small because the β particles are absorbed by glass and dissipated in a short distance in air. However, contamination of the skin or ingestion is particularly undesirable, because the radioactive amino acid will be incorporated into cells, where the $\beta$ particles can cause genetic damage.

- Harvest cells (5 to 10 million).
- Wash twice in PBS.
- Wash once in methionine-free medium (serum-free or with 10% dialyzed FCS).
- Resuspend in methionine-free medium at $10^6$/mℓ and place in incubator for 1 hr (to utilize intracellular pool of methionine).
- Add $^{35}$S-methionine (100 μCi for $10^7$ cells).
- Incubate 4 to 6 hr.
- Centrifuge, wash once in PBS.
- Make extract as described in the previous section and determine %TCA-precipitable counts.

iii. Notes

Whereas iodination by the Iodo-gen procedure will label principally protein exposed on the outer surface of the cell, methionine labels internal constituents, including intermediate forms which are processed before being expressed on the surface, and proteins which are not expressed on the membrane.

Mixtures of amino acids labeled with $^{14}$C or $^3$H may be used instead of methionine. However, because the specific activity of $^{35}$S-methionine is much higher than that of $^3$H or $^{14}$C amino acids, methionine will give the best labeling unless it is present in very small amounts in the protein of interest.

Some idea of the rate of turnover of the protein is helpful (see Chapter 5, Section II.D.2), because some proteins turn over slowly and the culture period may need to be extended.

$^{35}$S is a $\beta$-emitter and must be counted in a liquid scintillation counter after addition of scintillation cocktail. Detection of the antigen after electrophoresis is best done by fluorography (see Section 4.b.iv).

*2. Immune Precipitation*

*a. Membrane Extracts*

i. Materials

- Centrifuge to wash precipitates rapidly. We use the Beckman Microfuge 12 (Beckman Instruments).
- Beckman Microfuge 1.5-μℓ polypropylene tubes or equivalent. Polypropylene is preferred to polystyrene because it absorbs less protein.
- *Staphylococcus aureus* fixed organisms. This should be a strain which expresses protein A on the surface (Cowan 1 strain). The formalin-fixed organisms are available commercially from a number of suppliers, for example, Immunoprecipitin® (Bethesda Research Laboratories). The ability to absorb immunoglobulin seems to be unstable, and a new batch, or one which has been stored for a long period, should be checked. This is easily done if a quantitative assay for immunoglobulin (of a type which reacts with protein A) is available. We use human serum, because we have an automated routine assay for human immunoglobulin. Typically, 1 mℓ of a 10% v:v suspension of *S. aureus* will absorb 1 mg immunoglobulin. The suspension of fixed organisms should be stored frozen in aliquots of 0.5 mℓ. Before use, thaw out an aliquot and wash three times in lysis buffer, resuspending to a 10% v:v suspension.
- Washing buffers (these do not contain BSA, but may contain PMSF): (1) lysis

buffer with added NaCl to 0.5 *M*; (2) lysis buffer with added detergent — 0.1% SDS; (3) 0.1% NP40 in 10 m*M* tris, pH 7.4.

- Sample buffer: 10% SDS, 2 mℓ; glycerol, 1 mℓ, 1 *M* Tris, pH 6.8 (electrophoresis gel buffer — see Section 4.a), 0.8 mℓ; water to 10 mℓ; store frozen. For use thaw and add 5 mℓ bromophenol blue stock solution and, for samples that are to be reduced, 0.1 mℓ dithiothreitol stock solution per milliliter sample buffer.
- Bromophenol blue stock solution: 0.2%.
- Dithiothreitol (DTT) stock solution: 1 *M*, 154 mg/mℓ. Store frozen. WARNING: DTT is a toxic thiol with a strong smell. It should be handled in a fume cupboard.

ii. Method

- Preabsorb extract with 100 μℓ of *S. aureus* suspension per milliliter extract, for 30 min on ice. Spin in microfuge at 4000 × G for 2 min.
- Aliquot radiolabeled extract — 100 to 200 μℓ into Beckman tubes.
- Add monoclonal antibody — typically 20 μℓ ascitic fluid — and incubate on ice overnight.
- Add rabbit-antimouse immunoglobulin — typically 50 μℓ Dakopatts cat. #Z109. Mix and leave on ice 60 min.
- Add 50 μℓ *S. aureus* suspension. Mix and leave on ice for 60 min.
- Spin the suspension in the microfuge at 4000 × G for 2 min.
- Remove supernatant and wash the precipitate in the sequence of three washing buffers. Each wash consists of the following steps: carefully remove the supernatant, ensuring that the pellet is not disturbed, but all the liquid is removed. Add 1 mℓ cold wash buffer and resuspend the pellet thoroughly with a Pasteur pipette, ensuring that a uniform suspension is obtained. Clumps trap unwanted impurities. Place in the ice bath for 5 min. Centrifuge. Retain all supernatants and count 5-μℓ aliquots to follow the washing.
- After the final wash remove all the liquid and add 50 μℓ sample buffer. Resuspend the pellet thoroughly with a Pasteur pipette and transfer to a clean Beckman tube (because some radioactivity will have stuck to the tube during the washes). Take a 5-μℓ sample for gamma-counting and place the remainder on a bath of boiling water for 4 min. The tube cap may be perforated with a hypodermic needle (left in place during boiling) to prevent the cap from popping open.
- Centrifuge once more to pellet the *S. aureus* organisms. The sample is now ready for electrophoresis.

iii. Notes

Wash buffers — The rationale is to dissociate nonspecific complexes and leave the antigen-antibody complexes precipitated. The different washes (high salt, SDS, low salt) are to dissociate different nonspecific bonds. Many other washing buffers could be used (the ones quoted were in use in Dr. M. Crumpton's Laboratory when the author learned these techniques), and different buffers should be used if the final precipitate is still contaminated with coprecipitated protein, or if the antigen-antibody complex dissociates during washing. Monitoring the radioactivity of the washes helps to guard against the latter possibility, and the wash, if found to contain the antigen, can be electrophoresed. However, it would not be appropriate to place too much emphasis on the value of the counts in the final sample, since we do not know, until the autoradiograph is completed, whether they are concentrated into a single band or represent a generalized background.

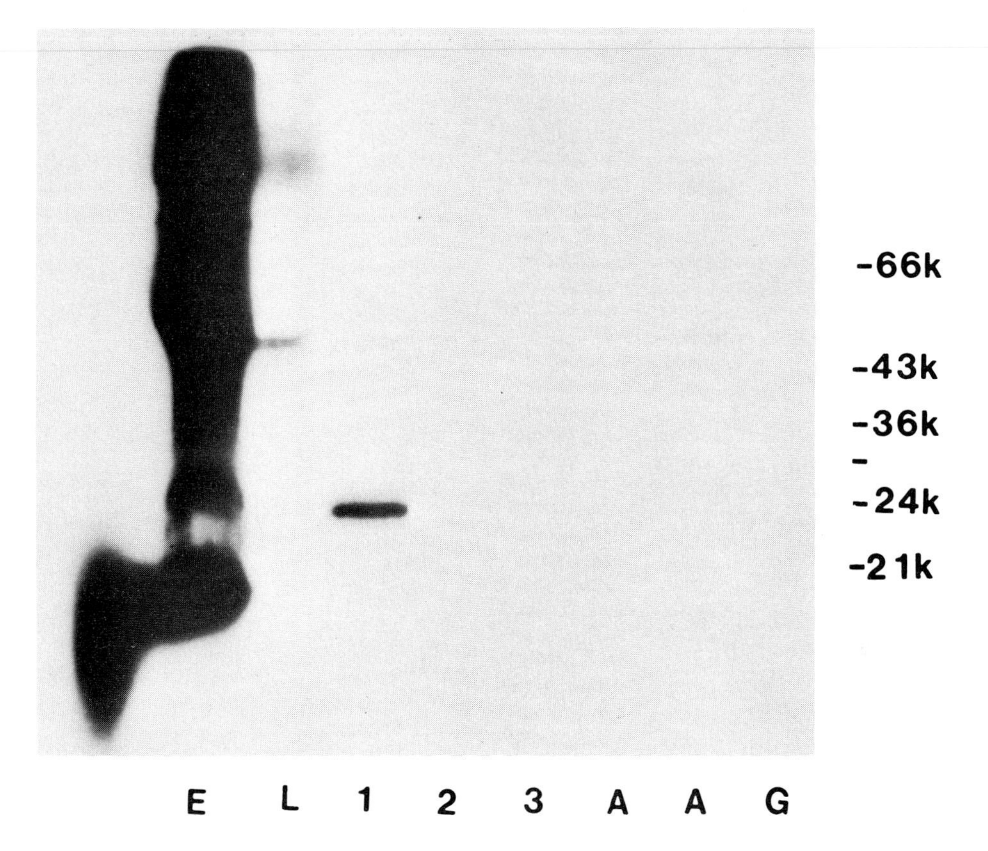


FIGURE 8. Fractions eluted from a column of monoclonal antibody FMC56. The antigen is eluted in fraction 1.

*b. Soluble Proteins*

The method described in the previous section may be used essentially unchanged for soluble, as opposed to solubilized membrane, antigens. Instead of starting with a detergent extract the protein, at a concentration of 1 to 10 mg/mℓ (depending on purity), is mixed with monoclonal antibody. The rest of the procedure does not need to be changed, although the wash buffers may be replaced by a simple PBS or tris buffer. Conditions which may dissociate immune complexes — high or low pH, chaotropic ions, certain detergents (see Section V) — should be avoided.

*3. Affinity Column*

Although precipitation as described above allows the characterization of some protein antigens, frequently the result is a series of bands of low intensity which do not allow the positive identification of the antigen. In this situation we and others have attempted to purify the antigen by affinity chromatography on a small column of the monoclonal antibody. The techniques involved are described in detail in Section V. The column method uses more antibody and probably takes longer, but offers two potential advantages over precipitation. First, since the only material eluted from the column should be derived from the antigen extract, there is no need to use radioactive isotope, provided a sensitive protein stain is used (silver stain, see below). Second, it is easier to try a series of washing and eluting buffers and run each one by electrophoresis. Figure 8 shows an example where the antigen was eluted in what was intended to be a washing buffer.

*4. Polyacrylamide Gel Electrophoresis*

*a. Electrophoresis*

Many different electrophoretic separation procedures have been described, with advantages in particular situations. The method described here is that of Laemmli,[18] which gives high resolution of proteins and glycoproteins and an indication of molecular weight (see discussion in Section B.2.a).

i. Materials

- Slab gel electrophoresis apparatus: may be made by a competent workshop and is available from several manufacturers — Bio-Rad, LKB, Pharmacia, and others. Requisites are ease of assembly and running, cooling system, reasonably priced and readily available replacement parts. The power pack must be capable of delivering 250 V at 35 mA.
- Buffers and gel chemicals: these must be of high purity since impurities interfere with the separation and staining. Companies which specialize in electrophoresis, such as Bio-Rad, provide electrophoresis-grade chemicals. If the gel is to be silver-stained, all solutions have to be filtered through a 0.45- or 0.22-μm membrane to reduce background staining. If solutions are stored, the pH should be checked and adjusted before use, since the quality of the separation depends critically on pH differences between the two gels and the running buffer.
- Acrylamide stock: WARNING: acrylamide is a neurotoxin and can diffuse through the skin. Acrylamide and solutions containing acrylamide should be handled with gloves; 30 g acrylamide + 0.8 g bis-acrylamide (*N,N′*-methylenebisacrylamide) dissolved in 100 mℓ water. Store at 4°C in a foil-wrapped bottle, and use within 30 days.
- Tris buffers for gel preparation: dissolve 12.11 g tris base in 80 mℓ water, add 5 *M* HCl to adjust pH to 6.8 (stacking gel buffer) or 8.8 (running gel buffer), and make up to 100 mℓ.
- Sodium dodecyl sulfate (SDS), 10 g in 100 mℓ water. This solution gels in the refrigerator; redissolve by warming.
- Electrophoresis buffer: tris base, 3.03 g; glycine, 14.42 g; SDS, 1 g; make to 900 mℓ with distilled water, adjust pH to 8.3, and make up to 1 ℓ.
- Ammonium persulfate: 100 mg/mℓ, made up within 1 hr of use. This oxidizing agent initiates polymerization of the gel; if the solution is kept for too long, polymerization will be slow and uneven.
- TEMED: *N,N,N′,N′*-tetramethylethylenediamine (Bio-Rad)
- Molecular weight markers: ready-made mixtures are available from several companies (Sigma, Pharmacia, Bio-Rad). An advantage in making up mixtures from individual proteins is that the purity and electrophoretic mobility of each marker may be checked. Useful markers include: Lactalbumin (MW 14,400); trypsin inhibitor (20,100); carbonic anhydrase (29,000); ovalbumin (45,000); bovine serum albumin (66,000); phosphorylase B (92,500); β-galactosidase (116,000); and myosin (205,000). The proteins are made up in sample buffer at concentrations ranging from 0.25 to 1 mg/mℓ and 20 μℓ of the mixture is applied per lane. An example of a marker run may be seen in Figure 6.

ii. Method

Set up the plates in the gel casting apparatus, according to the manufacturer's instructions. The plates must be thoroughly cleaned, by washing and then wiping with an alcohol-soaked clean cloth. The use of grease to prevent leaks should be avoided; grease inhibits polymerization. If leakage of the acrylamide solution occurs and cannot be prevented by working carefully, use agarose gel to cover the outside of the plate at the base.

Make up the running gel. The volume required will depend on the dimensions of the gel. Volumes given below are for a 1.5-mm gel in the Bio-Rad apparatus. The gel bed should be as long as possible, but should allow for about 1 cm of stacking gel between the bottom of the sample well and the top of the running gel. The gel porosity selected will depend on the anticipated molecular weight of the proteins of interest; 5% gels are needed to separate high molecular-weight proteins (up to 200,000), but are difficult to work with, being very fragile. At the other extreme, 10 or 12.5% gels may be used for low molecular weight proteins (down to 12,000) and are robust. Because the linear relationship between log (MW) and migration only holds over a limited range, use 12.5% gels up to 45,000 MW, 10% gels from 10,000 to 70,000 MW, and 5% gels from 25,000 to 200,000.

To make the gel, warm the solutions to room temperature and pipette 5, 7.5, 10, or 12.5 mℓ (corresponding to the % gel selected) of acrylamide stock solution into a clean conical flask; add 11.2 mℓ of tris pH 8.8 buffer (running gel buffer) and water to a final volume of 30 mℓ; mix and degas by connecting the flask to a vacuum line and swirling the mixture around until bubbles of gas are seen to form. Then add in sequence: SDS, 0.3 mℓ; ammonium persulfate, 0.1 mℓ; TEMED, 20 $\mu$ℓ. Mix and immediately pour the gel.

Carefully overlay the gel with a thin layer (2 to 3 mm) of water, avoiding mixing of the water with the acrylamide. The water layer is added to prevent direct contact between air and the acrylamide, since oxygen inhibits polymerization.

Allow to polymerize. If this takes more than 1 hr something is wrong with the mixture — the ammonium persulphate or TEMED are inactive, the mixture is not adequately degassed, or the temperature is too low.

Make up the stacking gel: use a 3% stacking gel for 5 and 7.5% running gels, and a 5% stacking gel for 10 and 12.5% running gels.

Mix the following: acrylamide stock, 1.67 mℓ (5% gel) or 1.0 mℓ (3% gel); stacking gel buffer (tris pH 6.8), 1.25 mℓ; water to 10 mℓ (7.03 mℓ for 5% gel; 7.7 mℓ for 3% gel). Degas, then add: SDS, 0.1 mℓ; ammonium persulfate, 50 $\mu$ℓ; TEMED, 20 $\mu$ℓ. Remove the water from the top of the running gel; pour the stacking gel and insert the "comb" which forms the sample wells, ensuring that air bubbles are not trapped in the process. Allow the gel to form (this should not take longer than 1 hr).

Remove the "comb" gently, so as not to disturb the gel which now separates the sample wells. Fill the sample wells with electrophoresis buffer and underlay the samples (5 to 50 $\mu$ℓ) carefully, avoiding mixing. The sample buffer contains glycerol to make it denser than the electrophoresis buffer, but a steady hand is still required, and air bubbles can disturb the sample layer. Samples may be applied with calibrated disposable capillary tubes or with a microsyringe. If a microsyringe is used it must be washed five times between samples to avoid cross-contamination. Without wasting any time, place the gel in the apparatus and switch on the current. Ensure the correct polarity is used; proteins in SDS are negatively charged and will migrate towards the anode (+).

Use 50 V potential difference until the bands enter the running gel, then 10 to 15 V/cm gel length. When the marker band nears the bottom of the gel stop the electrophoresis and remove the gel for staining. Using a water-cooled gel assembly with running gels of 15 cm length, the run should take about 5 hr. The faster it is run the sharper the bands will be, provided the cooling system is adequate.

### *b. Staining and Antigen Detection Methods*

Protein staining is normally carried out with Coomassie blue, unless higher sensitivity is required, in which case the silver stain is used. If the gel has radioactive bands it is stained with Coomassie blue before drying for autoradiography or fluorography. The Coomassie stain is in a solution which also fixes the protein. Gels which are to be blotted are not stained before blotting, since the protein would be fixed, but may be stained after blotting.

#### i. Coomassie Blue Staining

Coomassie brilliant blue R250 (Bio-Rad) (color index #42660) or brilliant blue R (Sigma Catalog #B0630): weigh out 1.25 g and dissolve in 300 mℓ methanol. Filter through a paper filter on a funnel. Add 70 mℓ acetic acid (glacial) and make up to 1 ℓ with water.

Place the gel "sandwich" in the stain and, wearing gloves, gently pry apart the glass plates. The gel should float off. Leave in the stain for 1 to 2 hr, then transfer to fresh stain and leave overnight at room temperature.

Pour off the stain and replace it with destaining solution: 30% methanol, 7% acetic acid in water. Leave at room temperature on a rocking platform, changing the destaining fluid several times, until the gel is destained. Usually this will take four to six changes and 6 to 8 hr, but it may be necessary to destain overnight.

#### ii. Silver Stain

The method described here is based on that of Morrissey.[19] All solutions should be filtered through a 0.2-μm membrane to reduce background staining. Remove the gel from the apparatus and fix in 50% methanol/10% acetic acid for 30 min, followed by 5% methanol/7% acetic acid for a further 30 min. Transfer the gel to 10% glutaraldehyde, and add solid $NaHCO_3$ to neutralize the acid, until further addition does not result in effervescence. Leave the gel in the glutaraldehyde fixative for 30 min, then transfer to water. Soak the gel overnight in an excess of distilled water. Wash in a fresh lot of water for 30 min, and soak in water containing 5 μg/mℓ dithiothreitol for 30 min. The dithiothreitol may be kept frozen as a 100 × stock solution, but the working solution should be made by diluting into filtered water and used immediately. Soak gel in 0.1% $AgNO_3$ solution for 30 min. Rinse once in distilled water. The gel is now ready for developing. The developer consists of 50 μℓ 37% formaldehyde (i.e. undiluted formalin) in 100 mℓ of 3% $Na_2CO_3$. Developer is added to the gel, which is kept moving in the dish and observed. After about 1 min some darkening will be seen, and the developer can be replaced with fresh developer. The image will begin to intensify, and development must be stopped before the background gets too dark. Further changes of developer may be needed, and the results improve with practice. Development is stopped by adding citric acid (48.3 g/100 mℓ) to neutralize the $Na_2CO_3$ in the developer. The amount to be added will depend on the volume of developer used and should be determined beforehand. If the liquid is still alkaline development will continue; if it is acid some fading will occur. Remove the neutralized developer and wash in water. If the gel is to be photographed this should be done sooner rather than later, since the background may continue to darken.

#### iii. Autoradiography for $^{125}I$

The gel must first be dried onto a backing of filter paper. The destained gel is transferred to a solution of 10% acetic acid, 1% glycerol in water. This solution promotes even drying and reduces cracking. After 1 to 2 hr the gel is floated onto filter paper (2 sheets of Whatman 3MM) and placed on the metal grid of a gel dryer (Bio-Rad). It is overlaid with a sheet of thin plastic (such as cling-film), taking care to avoid air bubbles and stretching of the gel. A sheet of Mylar (supplied with the gel dryer) is placed over the gel and the rubber cover of the gel dryer is put in place. The dryer is attached to a vacuum pump, the vacuum established by stretching the rubber cover over the gel, and the heater turned on. Drying time depends on the gel and the vacuum, but is generally 2 to 3 hr with heat followed by 30 to 60 min with the heater off. If the vacuum fails during drying the gel will break up. A good venturi pump will dry gels satisfactorily, but if the water pressure is inadequate a vacuum pump may be used. If a vacuum pump is used, a trap must be included in the line to prevent contamination of the pump oil and corrosion of the pump by acetic acid. The amount of acetic acid to be removed may be reduced by washing the gel in 1% glycerol in water for 1 hr immediately before drying.

The dry gel can now be placed in contact with film to detect radioactive bands. Sensitivity is much improved by using scintillation screens which convert the gamma radiation into light photons. The system consists of: the gel, a sheet of paper to prevent contamination of the scintillation screens, and a "sandwich" consisting of the film between two scintillation screens. The screens have their scintillant faces inwards towards the film. The film, thus, receives photons from both sides, and to take advantage of this double-sided film should be used.

The whole assembly is placed between a pair of metal plates and clamped together tightly, ensuring close contact between gel, screens, and film. Special cassettes are available for this purpose, but a suitable assembly may be made with strong "bulldog" clips and metal sheets cut to size.

The assembly is placed in a −70°C freezer. The nascent silver grains formed by exposure to photons decay at higher temperatures. The freezer should not contain other sources of radiation.

Exposures vary from 1 day to 2 weeks, depending on how hot the sample is and on the type of film and screens used. As a rough guide, if the sample contains a total of 1000 cpm a clear pattern will be seen after a week; a sample with 4000 cpm will give a strong band after a week and a clear one after only 24 hr. However, if the counts represent multiple "background" bands rather than a single protein, a strong band will not be seen. The film should be a fast X-ray film, double-sided, such as Kodak XRP5. The screens should produce light to which the film is sensitive; not all screen-film combinations are suitable. We use Ilford Fast Tungstate screens. The sensitivity of the film can be increased by prefogging.[20] This process will need to be calibrated for the light source and filter used.

iv. Fluorography for $^{35}S$ or $^{3}H$

Stain the gel (Coomassie blue) and destain. Soak the gel in two changes of DMSO (200 mℓ; 30 min at room temperature). Soak for 3 hr in 100 mℓ DMSO containing PPO 22.2%, on a rocking platform. PPO (2,5-diphenyloxazole) is a scintillant; it converts the energy of $\beta$ particles into light photons. Soak in two washes (200 mℓ or more) of 10% acetic acid, 30 min each wash, at room temperature. If the smell of DMSO is still detectable, soak again in acetic acid. Place the gel in drying solution, and dry the gel (see previous section).

Expose in direct contact with X-ray film, at −70°C, under pressure (see previous section for details). The film may be prefogged for greater sensitivity. Note that there must be nothing between the film and the gels, since the emission being detected is light rather than penetrating radiation.

Fluorography is about ten times more sensitive than autoradiography for $^{35}S$. Scintillation cocktails suitable for fluorography may be obtained ready-made (ENHANCE, New England Nuclear). A cheaper alternative is the use of salicylate.[21]

v. Electrophoretic Blotting

Materials include:

- Electrophoretic transfer apparatus (Bio-Rad Trans-Blot or equivalent). Power-pack for blotting (capable of delivering 200 mA at 70 V). For some applications high field strengths are needed, and it may be worth purchasing a larger power supply at the outset (see Bio-Rad catalog).
- Nitrocellulose: 0.45- or 0.22-$\mu$m pore size (available from Bio-Rad). The 0.45-$\mu$m size is used for most purposes, but low molecular-weight proteins may be better absorbed by the 0.22-$\mu$m nitrocellulose.
- Transfer buffer: 3.025 g tris, 14.4 g glycine, and 200 m$\ell$ methanol; make up to 1 $\ell$ with water and adjust pH to 8.3.
- Blocking solution: 6% suspension of skim milk powder in PBS, centrifuged at 5000 × G for 15 min and filtered through a 0.8-$\mu$m followed by a 0.45-$\mu$m membrane. See Section II.C.4.b for alternative blocking solutions.
- Washing solution: PBS/Tween 20, 0.5 m$\ell$ Tween 20 (BDH) in 1 $\ell$ PBS
- Reaction buffer: 6.05 g tris, 8.76 g NaCl, 1.86 g EDTA, 2.5 g BSA, 0.5 m$\ell$ NP40; make up to 1 $\ell$ with water and adjust pH to 7.4.
- Detection system for mouse immunoglobulin: see Section II.C.

Cut the nitrocellulose sheet, wearing gloves to avoid putting fingerprints on the nitrocellulose. Soak the nitrocellulose in transfer buffer. Remove the gel from the plates and place it in the transfer buffer. Leave it soaking in transfer buffer for 30 min. Meanwhile, prepare the blotting assembly consisting of, in sequence: perforated plastic support, spongy support (Scotchbrite® pad cut to size), gel, nitrocellulose sheet, filter paper pad (Whatman 3MM), spongy support (Scotchbrite® pad), and perforated plastic support.

Assemble the components in transfer buffer, getting rid of air bubbles in the process. Place the assembly in the electrophoretic transfer apparatus. Ensure the polarity is correct: at the pH of the transfer buffer most proteins will be negatively charged and will move towards the anode (+). Thus, the nitrocellulose should be anodal to the gel. Note that some proteins may move the other way, especially if the gel is not an SDS gel. If the antigen of interest does not blot, it may be basic; try transfer with the polarity reversed.

Run the transfer apparatus in the cold-room overnight at 100 mA constant current. Remove the assembly and wash the nitrocellulose in washing solution five times (2 min/wash) and once in PBS. At this stage it will be necessary to cut strips for reaction with different monoclonal antibodies, and this must be done on the basis of careful matching of the nitrocellulose to the wells in the gel.

Protein may be stained in 0.1% amido black in 45% methanol/7% acetic acid for 1 min, and the background destained in 5% methanol/7% acetic acid.

Antigens are detected by soaking with monoclonal antibody followed by detection reagents for mouse immunoglobulin, as described for dot-blotting in Section II.C. The sensitivities of the $^{125}$I and immunoenzymatic detection systems are about equivalent, and proteins can be detected in the 10-pg to 10-mg range, depending on the details of the detection test. The amount to be loaded depends on the composition of the sample. As a rule of thumb, the sample load should be 50 to 100 $\mu$g protein or, for cell extracts, the material from $2 \times 10^6$ cells.

## V. PURIFICATION: AFFINITY SEPARATION BASED ON MONOCLONAL ANTIBODY

### A. Introduction

The potential value of affinity purification using monoclonal antibodies would be difficult to overstate. Although the technical difficulties have deterred some workers, there are now enough published examples to illustrate the potential of the method. One example is shown in Figure 8, where a single pass through a column of the monoclonal antibody FMC56 results in the isolation of the antigen from a very crude and complex cell extract. The same purification by standard biochemical procedures would require the application of multiple separation steps based on different physical properties of the antigen, and would require a larger amount of starting material. The power of immunoaffinity as a separation technique justifies expending some effort to master the technical difficulties.

The major source of technical difficulty is simply that it is not possible to define a set of conditions which will work every time, because the nature and affinity of the interaction between antigen and antibody varies from application to application. In view of the variation in conditions required to carry out affinity purification successfully, this section consists of a detailed discussion of the variables involved. No single method is described in recipe form, but references are given to a number of published examples.

The basic method is to immobilize the antibody on a solid matrix, and use this immunoabsorbent as a column for affinity chromatography. There are a few shortcuts, such as using the immunoabsorbent in a batch mode. The most widely used shortcut is to use the antibody, in solution, to precipitate the antigen. This has been described separately (Section IV.C.2), since it is primarily used as a method for the identification of the antigen, rather than for purification.

As is the case for any purification procedure, an assay for the material to be purified is helpful in setting up the purification, and essential if the purification is to be carried out optimally. There are many examples of successful purification carried out without an assay for the material to be purified, but if the attempt at purification runs into difficulty, it is well worth setting up an assay. Dot-blotting or ELISA (Section II) provide suitable assays, since antibody is available against the antigen being purified. Alternatively, if the antigen has a characteristic electrophoretic mobility, fractions may be run in electrophoresis gels.

The factors to be considered in immunoaffinity purification by the column procedure are as follows:

1. Antigen source.
2. Choice of antibody.
3. Choice of solid matrix.
4. Choice of coupling method.
5. Column parameters.
6. Wash buffers.
7. Elution buffer.
8. Regeneration and reuse of column.

## B. Detailed Consideration of Variables

### *1. Antigen Source*

The mixture should contain as high a proportion of antigen as possible, and some preliminary purification by standard biochemical means may be worthwhile. However, immunoaffinity purification can yield very high purity from a crude extract. The extract will normally be at a neutral pH and should be in a buffer which allows antigen-antibody interaction. Thus, the dissociating factors used to elute the antigen from the column (see below) should be avoided in the original mixture. Extracts of membrane proteins will often contain detergent to keep the lipophilic proteins in solution, and the detergent may inhibit the binding of the antigen to the antibody. Detergents differ in their effects on binding affinity and it is worth experimenting with different detergents. The antigen solution should be filtered and centrifuged to remove insoluble fats and particles, which will clog the column.

*2. Choice of Antibody*

If several monoclonal antibodies against the same antigen are available, a choice may be made on the basis of the affinity of the antibody for the antigen, as well as any cross reactivity with other components of the antigen mixture. The higher the affinity the better the antibody will be able to pull the antigen out from the mixture and hold it while unbound components are washed off. However, elution of the antigen from the column under conditions which do not damage either the antigen or the antibody may be difficult if the affinity is too high. At present such a choice can only be made empirically, taking into account the stability of the antigen and antibody to the various alternative eluting buffers. The effective affinity depends on the column parameters; a low affinity antibody coupled to a column at high density may have, within limits, the same effective affinity as an antibody of higher affinity coupled at lower density. Furthermore, the size of the column may be important if the affinity of binding is such that the antigen is being absorbed and desorbed in the starting buffer. The important kinetic properties of the antibody-antigen interaction go beyond the affinity constant; two antibodies with similar affinity constants can have different rates of association and dissociation, and binding may or may not be highly dependent on bi- or multivalent interaction. The importance of determining the suitability of an antibody experimentally is emphasized by Parham,[22] who obtained essentially similar performance in columns made from two monoclonal antibodies with very different kinetic properties.

The antibody should be purified from ascites before coupling it to the support matrix. The amount needed will, of course, depend on the object of the project; as a guide 10 mg should enable the preparation of a 5-mℓ column with a capacity to purify roughly 1 mg antigen/run.

*3. Choice of Affinity Matrix*

A number of affinity supports are available commercially (Sepharose from Pharmacia; Affi-gel from Bio-Rad Laboratories). Differences include the composition of the solid support, the addition of "spacer" arms which keep the coupled antibody away from the solid support matrix (and so improve access), and the chemical groups provided for coupling the antibody. The ideal support matrix binds the antibody irreversibly (no leakage) and does not bind proteins nonspecifically. Nonspecific binding of proteins is more likely if the support contains charged groups or if charges are introduced during coupling, and certain spacer arms can contribute to hydrophobic interaction with protein. Nevertheless, a number of commercially available materials are effective and the choice is largely a matter of personal preference. The reader wishing to embark on an extensive series of purifications should compare the different support materials in actual use in the system under study.

*4. Choice of Coupling Method*

Some of the support materials are available for use with several alternative coupling methods. Again, there is limited experimental evidence supporting the superiority of one coupling method over another, although a comparative study (not with monoclonal antibodies) has shown that immunoabsorbents based on ether linkages (derived from epoxy-derivatized agarose) are more stable than those prepared from cyanogen bromide-activated agarose or thiol-based materials.[23] It is advantageous to understand the chemistry of the coupling reaction,[24,25] since subsequent decisions, such as the choice of buffers, should be based on this information. Coupling reactions generally involve a nucleophilic displacement reaction, in which electron-rich atoms (N or S) on the protein react with the derivatized matrix. Ammonium and tris salts as well as azide inhibit the reaction and should not be included in the coupling buffer. The protein may be displaced, after coupling, by sodium azide or by ammonium and tris ions, which should be avoided. The isourea link produced from CNBr-derivatized gels is more susceptible to displacement reactions than are amide (Affigel-10, Bio-Rad) and ether (epoxy-activated Sepharose, Pharmacia) linkages. Another consideration is that the formation of the isourea linkage introduces an extra charge, leading to an increased risk of nonspecific absorption. A new product from Pharmacia is Tresyl-activated Sepharose, which is said to produce more stable bonds, through amino and thiol groups.

The antibody may be bound directly to the matrix, or an antiimmunoglobulin (or Staphylococcal protein A) may be bound to the matrix, and the monoclonal antibody bound subsequently. If the second approach is used, it is advisable to stabilize the complex by chemical cross-linking, otherwise the monoclonal antibody may be eluted with the antigen (see for instance Schneider[26]).

In some cases the solid support matrix is available unmodified, and the derivatization is carried out in the laboratory. Methods are supplied by the manufacturers and in a variety of reviews (see, for example, Parham[22] for cyanogen bromide activation of CL-Sepharose 4B). Unless large-scale work is planned, it is more economical to buy ready-derivatized support material.

*5. Column Parameters*

The size of the column will depend on the amount of antibody bound and on the amount of antigen to be prepared, and must be determined by experiment. Typically, 2 to 3 mg of purified antibody can be bound per milliliter of support bed (Sepharose CL-4B, Pharmacia) and the column will have a capacity of 0.1 to 0.5 mg of antigen per milliliter of bed. If the column is larger than it needs to be, nonspecific absorption will increase and yields will decrease. If the antibody is bound by multiple links to the matrix, its binding capacity may be reduced because of its constrained conformation. It is, thus, probably better to have a slight surplus of antibody during the binding reaction.

Ideally, immunoaffinity purification involves the antigen being bound irreversibly in the starting and washing buffers and then dissociated by the elution buffer. However, in some instances the antigen is reversibly absorbed and desorbed in the starting buffer. The separation is then genuinely chromatographic, depending on the antigen being retarded by the column and the impurities coming out ahead of the antigen. Column size and shape are much more important in this situation. Conversely, if the antigen is bound strongly, the column dimensions are less important.

A typical column setup consists of a precolumn containing nonderivatized support gel, or gel coupled with normal mouse immunoglobulin (to remove material which binds nonspecifically). The antigen can be passed through this column directly to the affinity column and it is possible to run a series of affinity columns, to extract different antigens from a mixture, in series. The column should be precycled immediately before use, by washing with the elution buffer (to remove any impurity which may be eluted later and contaminate the antigen) and then equilibrated in the sample buffer. The column is run at 4°C, unless temperature differences are to be used as part of the elution protocol. A slow flow rate (say 20 to 30 mℓ/hr for a 5-mℓ column) is used, and the sample applied followed by five to ten column volumes of sample buffer. One milliliter fractions are collected, and all fractions are kept for subsequent analysis.

*6. Wash Buffers*

Having applied the antigen-containing mixture to the column, it is necessary to wash off unbound material. This may be done using the sample buffer, but it may be advantageous to use mildly dissociating conditions to remove nonspecifically bound material (i.e., material other than the antigen). This approach relies on the assumption that such binding (due to electrostatic interaction, hydrophobic interaction, and, possibly, cross reactivity with the antibody) will be of a lower affinity than antigen-antibody binding. Washing with additional buffers must be monitored to see if the antigen is eluted, and a sequence of washes suitable for the antibody and mixture in use can be arrived at empirically (see Chapter 6 in Goding[25]). The buffers used to wash precipitates (Section IV.C.2) may also be used to wash a column.

*7. Elution Buffer*

Once the antigen is bound to the antibody on the column and other materials have been washed off, the antigen must be eluted. This requires a change in conditions such that the antigen-antibody complex dissociates. In conventional affinity chromatography the complex may be dissociated by adding an excess of the column-bound ligand, and this could conceivably be used in immunoaffinity purification, by adding excess monoclonal antibody to the column. This would lead to the elution of antigen-antibody complex. This approach has not been used to the author's knowledge, and may be worth considering if the antigen is highly sensitive to the chemical-dissociating conditions which are commonly used. These involve high or low pH, addition of agents which reduce the polarity of the solvent (for example, dioxane) or detergents, or the use of high concentrations of salts. Certain ions are more conducive to dissociating complexes; these ions are referred to as "chaotropic". Thiocyanate is one of the most chaotropic anions and potassium, or, better still, lithium, are chaotropic cations. Thus, KSCN or LiSCN at high concentration (3 *M*) is often useful in eluting affinity columns. Detergents may also be effective, particularly for membrane proteins which are more soluble in the presence of detergent. Alkaline pH (11.5) seems to be particularly useful for membrane proteins (see Chapter 6 in Goding[25]).

For each particular system, the elution buffer will need to be worked out empirically. If the antigen is available pure or can be specifically identified, elution systems may be tested by working with small replicate aliquots of the immunoabsorbent in tubes and determining the elution under different conditions. Usually, the elution conditions will have to be determined by experiment with the column. The questions to be asked are

- Is the antigen eluted?
- Is the antigen damaged?
- Is the antibody damaged?

Damage to antigen and antibody can be reduced by rapid neutralization or dilution of the eluting buffer. The column should be washed immediately after use (see next section). The eluate may be collected into a high-molarity neutral buffer (e.g., 1 *M* tris, pH 7.4) or into a tube containing solid buffer salts. Alternatively, a pH-stat may be used to neutralize the eluate. Elution should be monitored (for example, by using radioactive antigen), but the antigen may be expected to elute in one to two column volumes, if the eluting buffer does not allow reassociation of the antigen to antibody. The effectiveness of the system will also depend on physical parameters. Higher tem-

Table 2
PUBLISHED EXAMPLES OF AFFINITY PURIFICATION OF ANTIGENS USING MONOCLONAL ANTIBODY COLUMNS

| Antigen | Elution buffer | Ref. |
|---|---|---|
| Human histocompatibility antigen (HLA) | 0.05 *M* Diethylamine-HCl, pH 11.5 | 22 |
| Mouse histocompatibility antigens (H2) | Several, using high pH, high salt, or detergent | 27 |
| Human leukemic cell membrane antigens | 0.05 *M* Diethylamine, pH 11.5, with 0.5% deoxycholate | 26 |
| Human platelet glycoproteins | Tris/EDTA/Triton × 100 at pH 7.4 for one column; glycine/Triton × 100 at pH 2.4 for another column | 28 |
| Rabbit microsomal cytochrome P-450 enzymes | Electrophoresis sample buffer (without dithiothreitol and marker dye) with added 2% deoxycholate | 29 |

perature favors dissociation, but will also accelerate any reactions which damage the antigen or antibody. Elution may be more effective if the immunoabsorbent is removed from the column and eluted in a larger volume, with mixing.

Table 2 lists some published elution conditions, together with references.

### *8. Regeneration and Reuse of the Column*

The aims of the study may be achieved with a single use of the column, but in most studies cost effectiveness will be greatly improved if the column can be reused, at least at few times. Elution conditions may adversely affect the capacity of the column; this is more likely if pH extremes are used to elute the antigen. The column should be restored to a neutral pH and chaotropic ions washed off as soon as possible by passing ten column volumes of neutral buffer through. The column should be stored in the cold, in buffer containing a bacteriostatic agent. The covalent bonds linking the antibody to the matrix are subject to hydrolysis and there will be some spontaneous loss of antigen. As mentioned earlier, this loss is greater with some types of bond than with others and can be reduced by introducing cross-links. Probably, the most important factor limiting the life of an immunoabsorbent column is contamination with particulate and fatty material from the antigen mixtures. It is important to centrifuge and filter the mixture before applying it to the column. Columns which are running slowly because of contamination with particulate material or fats may be improved by removing the immunoabsorbent from the column and washing it extensively, including a "defining" step, before repacking. With care, immunoabsorbents are stable for 1 year or longer and can be reused extensively.

## REFERENCES


1. Haber, E., Katus, H. A., Hurrell, J. G., Matsueda, G. R., Ehrlich, P., Zurawski, V., Jr., and Khaw, B.-A., Monoclonal antibodies specific for cardiac myosin: in vivo and in vitro diagnostic tools in myocardial infarction, in *Monoclonal Hybridoma Antibodies: Techniques and Applications,* Hurrell, J. G. R., Ed., CRC Press, Boca Raton, Fla., 1981, 91.
2. Ruoslahti, E., Votila, M., and Engvall, E., Radioimmunoassay of alpha-fetoprotein with polyclonal and monoclonal antibodies, in *Methods in Enzymology,* Vol. 84, Part D, Langone, J. J. and Vunakis, H. V., Eds., Academic Press, New York, 1982, 3.
3. Albrechtsen, M., Massaro, A., and Bock, E., Enzyme-linked immunosorbent assay for the human glial fibrillary acidic protein using a mouse monoclonal antibody, *J. Neurochem.,* 44, 560, 1985.

4. Bradley, L. A., Franco, E. L., and Reisner, H. M., Use of monoclonal antibodies in an enzyme immunoassay for factor VIII-related antigen, *Clin. Chem.*, 30, 87, 1984.
5. Zaane, D. V. and Jzerman, J. I., Monoclonal antibodies against bovine immunoglobulins and their use in isotype-specific ELISAs for rotavirus antibody, *J. Immunol. Methods*, 72, 427, 1984.
6. Teitsson, I. and Valdimarsson, H., Use of monoclonal antibodies and $F(ab^1)_2$ enzyme conjugates in ELISA for IgM, IgA and IgG rheumatoid factors, *J. Immunol. Methods*, 71, 149, 1984.
7. Brock, D. J. H., Barron, L., and Bedgood, D., and Hayward, C., Prospective prenatal diagnosis of cystic fibrosis, *Lancet*, i, 1175, 1985.
8. Harris, H., Monoclonal antibodies to enzymes, in *Monoclonal Antibodies and Functional Cell Lines. Progress and Applications*, Kennett, R. H., Bechtol, K. B., and McKearn, T. J., Eds., Plenum Press, New York, 1984, 33.
9. Zola, H., Moore, H. A., Hunter, I. K., and Bradley, J., Analysis of chemical and biochemical properties of membrane molecules in situ by analytical flow cytometry with monoclonal antibodies, *J. Immunol. Methods*, 74, 65, 1984.
10. Biochemical Society, Policy of the journal and instructions to authors. Policy and organization of the journal, *Biochem. J.*, 209, 1, 1983.
11. Towbin, H. and Gordon, J., Immunoblotting and dot immunobinding — current status and outlook, *J. Immunol. Methods*, 72, 313, 1984.
12. Lee, W. M. F., Westrick, M. A., and Macher, B. A., Neutral glycosphingolipids of human acute leukemias, *J. Biol. Chem.*, 257, 10090, 1982.
13. Hansson, G. C., Karlsson, K.-A., Larson, G., McKibbin, J. M., Blaszczyk, M., Heryln, M., Steplewski, Z., and Koprowski, H., Mouse monoclonal antibodies against human cancer cell line with specificities for blood group and related antigens. Characterization by antibody binding to glycosphingolipids in a chromatogram binding assay, *J. Biol. Chem.*, 258, 4091, 1983.
14. Towbin, H., Schoenenberger, C., Ball, R., Braun, D. G., and Rosenfelder, G., Glycosphingolipid-blotting: an immunological detection procedure after separation by thin layer chromatography, *J. Immunol. Methods*, 72, 471, 1984.
15. Gahmberg, C. G. and Andersson, L. C., Selective radioactive labeling of cell surface sialoglycoproteins by periodate-tritiated borohydride, *J. Biol. Chem.*, 252, 5888, 1977.
16. Hakomori, S., Tumor-associated carbohydrate antigens, in *Annual Review of Immunology*, Vol. 2, Paul, W. E., Fathman, C. G., and Metzger, H., Eds., Annual Reviews, Palo Alto, 1984, 103.
17. Bolton, A. E., *Radioiodination Techniques*, Review 18, Radiochemical Centre, Amersham, U.K., 1977.
18. Laemmli, U. K., Cleavage of structural proteins during assembly of the head of bacteriophage T4, *Nature (London)*, 227, 680, 1970.
19. Morrissey, J. H., Silver stain for proteins in polyacrylamide gels: a modified procedure with enhanced uniform sensitivity, *Anal. Biochem.*, 117, 307, 1981.
20. Laskey, R. A. and Mills, A. D., Quantitative film detection of $^3H$ and $^{14}C$ in polyacrylamide gels by fluorography, *Eur. J. Biochem.*, 56, 335, 1975.
21. Chamberlain, J. P., Fluorographic detection of radioactivity in polyacrylamide gels with the water-soluble fluor, sodium salicylate, *Anal. Biochem.*, 98, 132, 1979.
22. Parham, P., Monoclonal antibodies against HLA products and their use in immunoaffinity purification, in *Methods in Enzymology*, Vol. 92, Colowick, S. P. and Kaplan, N. O., Eds., Academic Press, Orlando, 1983, 110.
23. Zola, H., Immunoabsorbents prepared from cell membrane antigens: efficiency of incorporation and shedding of protein during use, *J. Immunol. Methods*, 21, 51, 1978.
24. Porath, J. and Axen, A., Immobilization of enzymes to agar, agarose, and sephadex supports, in *Methods in Enzymology*, Vol. 44, Mosbach, K., Ed., Academic Press, New York, 1976, 19.
25. Goding, J. W., *Monoclonal Antibodies: Principles and Practice*, Academic Press, London, 1983.
26. Schneider, C., Newman, R. A., Sutherland, D. R., Asser, U., and Greaves, M. F., A one-step purification of membrane proteins using a high efficiency immunomatrix, *J. Biol. Chem.*, 257, 10766, 1982.
27. Mescher, M. F., Stallcup, K. C., Sullivan, C. P., Turkewitz, A. P., and Herrmann, S. H., Purification of murine MHC antigens by monoclonal antibody affinity chromatography, in *Methods in Enzymology*, Vol. 92, Colowick, S. P. and Kaplan, N. O., Eds., Academic Press, Orlando, 1983, 86.
28. Berndt, M. C., Gregory, C., Kabral, A., Zola, H., and Castaldi, P. A., Purification and preliminary characterization of the human platelet membrane glycoprotein Ib complex, *Eur. J. Biochem.*, 151, 637, 1985.
29. Reubi, I., Griffin, K. I., Raucy, J. L., and Johnson, E. F., Three monoclonal antibodies to rabbit microsomal cytochrome P-450 1 recognize distinct epitopes that are shared to different degrees among other electrophoretic types of cytochrome P-450, *J. Biol. Chem.*, 259, 5887, 1984.

# FURTHER READING

## Immunoassay

Maggio, E. G., *Enzyme-Immunoassay,* CRC Press, Boca Raton, Fla., 1980.

Fellows, R. E. and Eisenbarth, G. S., *Monoclonal Antibodies in Endocrine Research,* Raven Press, New York, 1981.

## Antigen Characterization and Purification

Goding, J. W., *Monoclonal Antibodies: Principles and Practice,* Academic Press, London, 1983.

Hames, B. D. and Rickwood, D., *Gel Electrophoresis of Proteins: A Practical Approach,* IRL Press, London, 1981.

Johnstone, A. and Thorpe, R., *Immunochemistry in Practice,* Blackwell Scientific, Oxford, 1982.

Towbin, H. and Gordon, J., Immunoblotting and dot immunobinding — current status and outlook, *J. Immunol. Methods,* 72, 313, 1984.

Bers, G. and Garfin, D., Protein and nucleic acid blotting and immunobiochemical detection, *BioTechniques,* 3, 276, 1985.

Parham, P., Androlewicz, M. J., Brodsky, F. M., Holmes, N. J., and Ways, J. P., Monoclonal antibodies: purification, fragmentation and application to structural and functional studies of class I MHC antigens, *J. Immunol. Methods,* 53, 133, 1982.

Lefkovits, I. and Pernis, B., *Immunological Methods,* Vol. 2, Academic Press, New York, 1981.

Lefkovits, I. and Pernis, B., *Immunological Methods,* Academic Press, New York, 1979.

Chapter 7

# PROSPECTS, PROBLEMS, AND LIMITATIONS IN THE USE OF MONOCLONAL ANTIBODIES

## I. SCOPE OF THE CHAPTER

In the 10 years since a practical method was described for making monoclonal antibodies, the technique has been applied in nearly every field of biology and medicine. Monoclonal antibodies have revolutionized some of the areas in which they have been applied, by providing the reagents with which to analyze the underlying phenomena. The outstanding example is cellular immunology, which has seen the concepts of cellular interactions in the control of immune responses develop in detail and acceptance. The basic concepts existed well before the availability of monoclonal antibodies, but they were based on experiments which were open to criticism because of the lack of specificity of the reagents used. Thus, allo-antisera, used in dissecting cell cooperation in mice, were contaminated with antiviral antibodies, which may or may not have affected the results. In man, experiments on cell cooperation and interaction in the immune response were based on technically difficult in vitro assays which did not reproduce well in different laboratories.

Monoclonal antibodies have provided the probes for identifying and separating cells, enabling the dissection of the mixture of interacting cells which together make the ingredients of the immune response. Having provided the reagents to analyze the immune response at the cellular level, monoclonal antibodies are currently being used to analyze the immune response at a further level of resolution, the molecular level. Membrane receptors for stimulation or suppression are being analyzed and purified, and their functional sites probed with monoclonal antibodies. In this field monoclonal antibodies and probes for DNA and RNA are being used as complementary tools.

While immunology has been, perhaps inevitably, the first discipline to make extensive use of monoclonal antibodies, these reagents are being used in many other disciplines. Yet other areas of biology and medicine have barely been touched by the new technology, even areas which do utilize immunochemical techniques. For example, most radioimmunoassays for hormones are still carried out with polyclonal antisera.

The purpose of this chapter is to look critically at the existing and potential uses of monoclonal antibodies; to strike a cautionary note on the overenthusiastic use of monoclonal antibodies; to evaluate realistically the potential of monoclonal antibodies in some areas where the new reagents have not been embraced enthusiastically. In particular, the potential and limitations of monoclonal antibodies as diagnostic and therapeutic substances will be discussed, and the potential advantages of human hybridomas will be considered. No attempt is made to predict the value of monoclonal antibodies in areas outside the author's sphere of competence. The reader who is, for instance, a plant biochemist should be able to assess the potential and limitations of monoclonal antibodies in plant biochemistry by judicious extrapolation of the comments made in this chapter, and throughout the book.

## II. MONOCLONAL ANTIBODIES AS REPRODUCIBLE REAGENTS

One of the major strengths of the hybridoma technique is that it allows the preparation of essentially unlimited amounts of reagent of constant quality. As a consequence, two different laboratories, for instance, laboratories in different countries, can utilize the same reagent for an assay and, thus, eliminate one of the major sources of variation. Within the same laboratory, the reproducibility of the reagent eliminates batch-to-batch variation and, thus, reduces interassay variation.

All this is correct in principle, but is it true in practice? The answer is that there is batch-to-batch variation, and a sample sent from one laboratory to another may not perform as well in the second laboratory. The variations are much smaller and easier to control than in conventional antisera, but they should not be ignored. The danger is the belief that monoclonal antibodies are perfect, invariant reagents. This belief leads to the omission of quality control measures and a lack of good results.

The causes of variation have been discussed in some detail elsewhere in this book. There are practical problems, such as instability of immunoglobulin on storage or in transit, variation from mouse to mouse in ascites Ig production, accidental cross-contamination, or mislabeling. There are more insiduous problems such as gradual loss of activity through the growth of a nonsecreting mutant clone. In vivo, the immunoglobulin variable region genes are subject to a remarkably high frequency of mutations (1/1000 per base pair per generation — $10^6$ times higher than the spontaneous rate for other genes[1]). Whether the rate of mutation is as high in hybridoma cells is not known.

The consequences of these considerations are obvious. Monoclonal antibodies provide excellent reagents, but only if adequate quality control measures are applied.

The exact nature of the quality control tests required depends to a large degree on the use of the antibody. For many purposes we have used undiluted tissue culture supernatant, stored in aliquots at −20°C, without any check on the Ig concentration. However, we always check activity. If ascites fluid is used it must be titrated to find the appropriate concentration; ascites from littermate mice given the same inoculum of cells can have quite different Ig concentrations. Material obtained from another laboratory should be checked for specificity and titer.

Such quality control measures are rather obvious and not difficult to carry out. The need for some control cannot be overemphasized. As both producers and users of monoclonal antibodies, we have experienced inactive batches from well-respected commercial suppliers, and the embarassment of having sent out monoclonal antibody which, from its reactivity as reported by the recipient, must have been mislabeled before dispatch.

The International Workshops on Leukocyte Antigens, discussed previously (Chapter 1, Section V), have illustrated the fact that different laboratories can obtain different answers with the same antibody. This may be attributed, in part, to inexperience or incompetence in a few laboratories, but some monoclonal antibodies are of such low reactivity that they will not give reproducible results when used in different laboratories. A weak reagent will be more subject than a strong one to differences in technique and in test interpretation. The author has several antibodies which perform reasonably well when used in indirect immunofluorescence by people who are accustomed to them, but would not perform well in a workshop. Such antibodies are simply not sufficiently robust in routine use to be good reagents. The same reagent may behave satisfactorily if used in an assay system with greater sensitivity or less subjectivity (for instance, flow cytometry rather than fluorescence microscopy).

## III. COMPARISON OF DIFFERENT MONOCLONAL ANTIBODIES FOR THE SAME USE

In some applications the user may have a choice between several antibodies with the same specificity. The choice may be a simple matter of comparing price and performance (the antibodies may differ in affinity or subclass). Purified reagents are preferable to unfractionated ascites, because there is a smaller risk of variation in antibody titer,

and (properly) freeze-dried material is preferable to liquid reagent, because the antibody will have a longer shelf life and be more stable in transit.

These matters are fairly obvious, but antibodies which have the same apparent specificity can react quite differently when examined under certain conditions. Two antibodies specific for the same cell may react with different molecules on the cell, and these antigens may be differentially expressed by cells at different stages of maturation. In this situation some cells would react with both antibodies, while other cells would react with only one of the antibodies. Two antibodies against the same antigen may be directed against different epitopes of the antigen and, consequently, have different functional effects. Differences in function can also result from differences in the Fc end of the molecule.

These several points can all be illustrated by considering monoclonal antibodies against human T lymphocytes. Several antibodies are available against the T3 (CD3) and T11 (CD2) antigens. In normal peripheral blood all these antibodies would be regarded as specific for T cells and would give essentially the same result. However, early T cells and many T-cell acute lymphoblastic leukemias would react with the CD2, but not the CD3 antibodies. Among the CD2 antibodies, reactivity can be detected against three distinct epitopes on the molecule. One of these epitopes is only detectable when the cell has been activated. Antibodies to one of these epitopes block the sheep erythrocyte rosette reaction; antibodies to the other two do not. Antibodies to CD3 are mitogenic for T cells, in a reaction which requires the participation of monocytes. The monocytes of some donors will only support the mitogenic effects of CD3 antibodies if these antibodies are of a particular Ig subclass.

Thus, the specificity of a monoclonal antibody can be considered at several levels — cell, molecule, epitope. The properties of the antibody, particularly the effects it may have on cell function, may depend on the epitope specificity of the antibody, as well as its affinity for the epitope and the Ig subclass of the antibody.

## IV. DIAGNOSTIC REAGENTS

If a reagent is to be used for the diagnosis or monitoring of human or animal disease a high level of reliability is required, since decisions on treatment will depend on the results of the test. In some countries strict government standards must be met before a reagent can be used for diagnosis; nevertheless, reagents not registered for this purpose will be used in a less formal way to derive information about the patient's condition. This is true, particularly with new reagents and tests which are still under evaluation. For example, the various markers for lymphocyte subpopulations were in extensive use before they received official recognition as diagnostic reagents. For the purposes of this section, "diagnostic reagents" includes any reagent used to obtain information about a patient's condition.

The comments made in the last two sections apply with extra weight to diagnostic, as distinct from research use of monoclonal antibodies. These reagents do vary from batch to batch and do lose activity on storage or in transit. Quality control procedures are, thus, essential, as they are in any other diagnostic assay. Furthermore, as indicated in Section III, reagents with the same apparent specificity may only be equivalent in the assay in which they were originally compared; their use in a different procedure or on a different tissue or sample type may reveal differences.

An additional problem applies to diagnostic use of procedures originally developed as research procedures. Diagnostic tests are carried out in a much larger number and broader range of institutions than are research experiments. Some of the hospitals carrying out diagnostic tests do not have the same range of skills and experience that are available in a research institution. This may mean that they will perform a well-established test more reproducibly than the research-based institution, but will have difficulties with new technologies. Reagents and tests which are to be used widely should be particularly robust, allowing considerable flexibility in methodology and interpretation without loss of validity.

Research laboratories developing new reagents and tests, and commercial suppliers, should be aware of this need for robust reagents and tests and should refrain from making available reagents which will only work in the hands of the specialist. Workshops such as the ones on Leukocyte Antigens, already referred to, are helpful in deciding which reagents are suitable; even here many of the participating laboratories are using techniques such as flow cytometry which are not universally available.

These problems notwithstanding, the successful use of monoclonal antibodies for diagnostic purposes has grown at an enormous rate over the last few years and is continuing to do so. The advantages of monoclonal antibodies — specificity and reliability — allow the use of assays which have an intrinsically lower degree of specificity; for example, immunoradiometric assays instead of competitive radioimmunoassays (see Chapter 6). These features of monoclonal antibodies also enable the routine identification of cells such as the T-cell subpopulations and certain malignant cells, which could previously only be identified by a small number of specialized laboratories. This has some associated dangers, which have been aired infrequently in the literature.[2-4] However, these dangers mean that we must curb our overenthusiasm, not abandon these useful tests.

The disadvantages of some monoclonal antibodies — excessive specificity leading to weak reactivity, failure to produce secondary phenomena such as precipitation, and susceptibility to false negatives due to mutations at the epitope detected — can be nullified by using appropriate mixtures of monoclonal antibodies. This trend is beginning to become apparent in commercial antibody products, as is the sale of kits which contain all the monoclonal antibodies and ancillary reagents required to make a particular differential diagnosis.

The use of monoclonal antibodies for diagnostic procedures which require injection of the antibody (diagnostic imaging) is considered in the next section, since the problems here are similar to those in therapeutic use.

## V. THERAPEUTIC USE OF MONOCLONAL ANTIBODIES

### A. Introduction

Antisera have long been used for therapy; indeed, one of the foundations of immunology was the development by von Behring in 1890 of antibody against diphtheria toxin, prepared in horses. The use of antisera in the treatment of infection decreased with the development of antibacterial drugs, but a consistent, if small, demand continued for animal antisera against bacterial toxins and snake venoms. Where possible, human serum (produced from healthy donors immunized with bacterial toxoids) was used, to reduce the chances of anaphylaxis, a life-threatening reaction to reexposure to animal protein. The use of animal sera increased again in the 1970s when antilymphocyte sera were developed to counter immunological rejection of organ transplants. The immunosuppressive nature of these sera, together with controlled administration, reduced the risk of anaphylaxis, and antilymphocyte serum remains an effective way of preventing or reversing rejection. The major problem has been the difficulty in demonstrating efficacy in a large, well-controlled trial, a difficulty due partly to the batch-batch variation of the sera. These polyclonal sera contain multiple antibodies against different cells of the immune system and different molecules on these cells, and the

Table 1
MECHANISMS OF ANTIBODY-MEDIATED CELL DESTRUCTION

| Mechanism | Accessory factors and cells |
|---|---|
| Complement-mediated cytotoxicity | Complement (serum enzymes) |
| Antibody-dependent cellular cytotoxicity | K cells (cells with Fc receptors) |
| Opsonization | Cells capable of ingesting antibody-coated particles — polymorphonuclear and mononuclear phagocytes |
| Drug delivery | None; the antibody is coupled to a cytotoxic drug or radioisotope |

malignant breast tissue, but not in other tissues. Such an antibody can, nevertheless, be used to localize metastases, since there cannot be normal breast tissue in other parts of the body.

When looking for tumor-specific antigens to serve as a target for monoclonal antibody therapy, malignant T and B lymphocytes are a special case. The antigen receptor of a particular clone of T or B cells is essentially unique to that clone and may, thus, be considered a tumor-specific marker. Therapy of B cell tumors using monoclonal antibodies against the immunoglobulin idiotype of the malignant clone is under active investigation.[7] This approach is likely to be limited because the monoclonal antibodies have to be prepared for each individual patient, unless antibodies to common idiotypes are used, in which case the response is not tumor-specific. A recent report[8] that, during therapy, a subclone of the tumor emerges, with a different idiotype, is not encouraging.

### 2. *Mechanisms of Action*

Having considered the various questions of safety (Section A), the investigator will need to consider questions relating to efficacy. In vitro studies provide only limited information on the way a monoclonal antibody will interact with its target cell in vivo. This is, in part, because there are several possible mechanisms by which the body, with the help of antibody, can eliminate a cell. In vitro models for these mechanisms can be developed, but these models do not tell us which mechanisms operate in vivo and which are quantitatively more important in vivo.

The mechanisms by which antibody can destroy cells, given the presence of suitable accessory factors, are listed in Table 1. Mouse monoclonal antibodies do not all fix human complement effectively, and at present the factors which determine whether or not a particular antibody will fix complement are not all known. It is doubtful whether antibody-dependent cellular cytotoxicity (ADCC) is a mechanism which operates at all in vivo. Lysis by ADCC requires only very small amounts of antibody on the target cell, but the lysis is very easily inhibitable by immune complexes. Opsonization is likely to be an important mechanism in vivo; it is known to be effective in removing bacteria.

The use of monoclonal antibodies to target cytotoxic drugs to react selectively with malignant cells provides an additional approach, likely to be much more powerful than the natural antibody-mediated cytotoxic mechanisms. Promising results have been reported in animal models and in clinical trials.[6] The questions of specificity and efficacy, discussed above, are no less important when the antibody is coupled to a potent toxin.

Antibodies against membrane antigens can induce the removal of the antigen from the cell surface, leaving the cell otherwise healthy and functional. This process is known as antigen modulation and is likely to limit the effectiveness of some monoclonal antibodies. Whether an antigen is modulated or not depends on the nature of the antigen, but different antibodies against the same antigen may behave differently mixture of specificities, affinities, and immunoglobulin subclasses is never the same in two bleeds from the same animal, let alone batches produced in different laboratories.

Monoclonal antibodies were quickly recognized as potential therapeutic agents for the control of graft rejection, offering greater specificity and reproducibility than conventional antisera. The possible use of monoclonal antibodies in the localization of tumors in vivo, and in tumor therapy, has also excited a great deal of enthusiasm. The use of monoclonal antibodies to neutralize bacterial or snake toxins has attracted less enthusiasm than their use in immunosuppression or cancer, in part because the existing antisera work adequately. Furthermore, a snake venom contains a multiplicity of pharmacologically active substances, and an effective antivenene must neutralize most or all of these.

When considering the use of a monoclonal antibody in vivo, questions of safety and efficacy must be examined thoroughly. Efficacy is discussed later, in the context of specific applications. There are three principal safety questions — anaphylaxis, cross reactivity, and the introduction of oncogenic material. Administration of monoclonal antibody may lead to an anaphylactic reaction to the monoclonal antibody itself. This is anticipated when using monoclonal antibody from a different species, and is one of the main reasons for wanting to use human monoclonal antibodies in patients (Section VI). It is interesting to note, however, that murine monoclonal antibodies are being used in clinical trials (see below) and reports of anaphylaxis are few.

Monoclonal antibodies injected into a patient may damage cells and organs which are not the intended target. Since cross reactions occur between unrelated tissues (B cells and renal tubules, T cells and nerve tissue); cross reactivity cannot be predicted and testing has to be exhaustive.

A separate consideration is the possible administration of virus or DNA which could conceivably be incorporated into the genome of the recipient. Monoclonal antibodies are produced by malignant cells; the mechanism by which these cells maintain their autonomy is not known. If a virus or provirus is involved in the autonomy of the myeloma line, the recipient may become infected. This may be considered a greater hazard with human hybridomas, since mouse viruses as less likely to infect humans. The use of monoclonal antibody prepared as ascitic fluid in mice carries an additional risk, since these mice harbor retroviruses.

### B. Monoclonal Antibodies in Cancer

There is a great deal of activity in the general area of tumor therapy with monoclonal antibodies. A review of the literature is outside the scope of this book, and a number of books and symposia has been devoted almost exclusively to this area.[5,6] This section will be confined to a few salient points intended to encourage a cautious and critical approach.

#### *1. The Question of Specificity*

There is a popular theory that tumors are immunogenic in their natural host. The immune response eliminates many tumors, according to this theory, and the growth of a tumor to a detectable size indicates a failure of the immune response to reject it. The theory is attractive, but much of the evidence supporting it is open to alternative interpretation. If we discard the idea that all tumors *must* have tumor-specific antigens, we must accept that it may not always be possible to make tumor-specific antibodies.

Having accepted this limitation, there remains a considerable potential for monoclonal antibodies in treatment and in diagnostic tumor imaging. For example, monoclonal antibodies originally thought to detect antigens specific for breast tumors were, subsequently, found to react with a tissue-specific antigen, found on normal as well as in modulation experiments. LeBien et al.[9] have suggested that high affinity antibody modulates, whereas lower affinity antibody does not. The presence of cells with Fc receptors can enhance modulation, and this must be taken into consideration when testing in vitro for the ability of an antibody to modulate the antigen.

*3. Diagnostic Tumor Imaging*

The detection of small tumors, including metastases, using monoclonal antibodies tagged with radioactive isotopes is under intensive investigation. A cautious appraisal suggests that, at present, the technique is not routinely capable of detecting small tumors which escape detection by alternative techniques. Further improvements will depend on solving a number of problems, relating both to the monoclonal antibodies and the imaging techniques.[10] The antibody uptake ratio (binding to the tumor/binding to surrounding tissue) needs to be improved, and this may be achieved by using multiple antibodies against tumor antigens to increase uptake by the tumor, and by using antibody fragments to reduce Fc-mediated uptake by surrounding tissue. Results are sufficiently encouraging[10] to merit the application of further resources in the area of tumor imaging.

C. Monoclonal Antibodies in Transplantation

*1. Organ Transplantation*

Organ grafts which are not syngeneic to the recipient are the target of an immune response, which leads to graft rejection. Monoclonal antibodies have contributed enormously to our understanding of the mechanisms of rejection, as they have to our understanding of other facets of the immune response. Monoclonal antibodies have also been seen as therapeutic reagents, ''clean'' successors to antilymphocyte serum. There have been several clinical trials using monoclonal antibodies either to prevent[11] or to reverse[12] graft rejection. Results have been promising, and at least in the case of one large trial the number of patients is adequate for evaluation. In this study[12] the murine monoclonal antibody OKT3, which reacts with mature human T lymphocytes, was administered to patients undergoing kidney graft rejection. The rejection was reversed in 94% of episodes, compared with 75% in the control group receiving conventional treatment. The benefits were longer term, since the group receiving OKT3 to reverse rejection had a 1-year graft survival rate of 62%, compared with 45% in the group on conventional therapy.

*2. Ex Vivo Treatment of Bone Marrow for Transplantation*

Bone marrow transplantation is carried out in two distinct sets of circumstances. In a number of diseases where the marrow fails to function normally (aplastic anemia and some forms of immune deficiency) functional reconstitution may be achieved with a marrow graft. In some forms of cancer the delivery of sufficient doses of cytotoxic therapy to eradicate the tumor will also destroy the bone marrow. Whether marrow failure is the primary disease or the result of treatment, a syngeneic (identical twin) marrow transplant is likely to restore function. For the majority of patients, syngeneic marrow is not available, and two immunological problems accompany nonsyngeneic transplantation. The marrow graft may be rejected, in much the same way as other organ grafts are rejected. Since the marrow is the source of immunocompetent cells, a marrow that is not rejected can mount an immunological attack on the recipient's tissue, a graft-vs.-host (GvH) reaction.

GvH reactions are initiated by T lymphocytes, and if these can be removed from the graft, GvH reactions are reduced or prevented. Trials on the use of monoclonal antibodies directed against T cells are underway in a number of centers.[13]

Patients who have a functioning marrow before treatment with cytotoxic drugs have an alternative option, autologous marrow reconstitution. In this approach, marrow is taken from the patient before ablative treatment, and cryopreserved. After treatment, an injection of the patient's own marrow cells can lead to reconstitution of marrow function. However, in some of the malignant conditions which might be treated in this way, the marrow acts as a reservoir of malignant cells (leukemia, neuroblastoma, certain lymphomas). In these patients the marrow must be "purged" of malignant cells. Clinical trials on purging of malignant cells with monoclonal antibodies by antibody-mediated cytotoxicity, using toxin-conjugated antibodies, and using magnetic micro-carriers coated with monoclonal antibodies are underway in leukemia[14,15] lymphoma and neuroblastoma.[16]

The specificity requirements for antibodies to be used in purging marrow, either of T cells or of malignant cells, are less stringent than for treatment of patients by direct administration of antibody. So long as the malignant cells (or GvH-inducing T cells) are destroyed and the essential stem cells of the marrow are spared, the antibodies may react with other cells in the marrow. Only minute quantities of antibody are transferred to the patient, so that reactivity with other tissues is relatively unlikely to cause problems.

## VI. HUMAN MONOCLONAL ANTIBODIES

The preparation of monoclonal antibodies by fusion of human B cells with human continuous cell lines has been attended with considerable technical difficulties, but the increasing number of successful reports indicates that these difficulties have been resolved, at least in part. These technical matters will not be discussed here; a number of references has been given in Chapter 3. The purpose of this section is to examine the situations in which human monoclonal antibodies may provide advantages over murine monoclonal antibodies.

One of the major difficulties anticipated, and to some extent experienced, with the in vivo use of murine monoclonal antibodies has been the immune response of the patient to the foreign immunoglobulin. This may reduce the efficacy of the administered antibody, by causing accelerated elimination. Furthermore, the immune response may take the form of anaphylaxis. The results published to date show little evidence of anaphylactic reactions; rather the immune response leads to accelerated clearing of the mouse monoclonal antibody and so reduces its efficacy. The patient's immune system may recognize idiotypic and allotypic determinants on human immunoglobulin as foreign, but the response is likely to be much weaker than that against immunoglobulin from a different species. Reports which indicate that the patient's immune response to murine monoclonal antibody is directed predominantly against idiotypic determinants have been seen as an indication that human monoclonal antibody may not be much more effective. However, it is very difficult to make reliable predictions in this context. It may be that the idiotypic determinants of mouse immunoglobulin are immunogenic because they are presented on a foreign carrier (mouse immunoglobulin), and that human idiotypic determinants would be much less immunogenic in man.

In diagnostic applications the human monoclonal antibodies have some potential advantages over murine reagents, but these are accompanied by two significant technical difficulties. The potential advantages derive from differences in the range of antigens recognized within and between species. This is discussed in more detail below. The major technical difficulty in making human monoclonal antibodies is that of immunization. For obvious reasons people cannot be immunized at will, and immune cells must be derived from a miscellany of sources — autoimmune patients, rash volunteers (usually the scientist who wants the monoclonal antibody), recent recipients of vaccine, recent victims of snakebite, etc. Alternatively, immune responses may be induced in vitro, a technique which has yet to be worked out satisfactorily with human cells. A related and subsidiary problem is that the cells commonly available from human donors are blood cells, and the numbers of B-cell blasts, the most probable participants in successful fusion, are much lower in blood than in spleen or lymph nodes. Nevertheless, a number of human monoclonal antibodies has been made from blood-borne B cells of volunteers. The second technical difficulty relates to the need to detect the binding of human immunoglobulin to human tissue, if the antibodies sought are against human tissue components. Detection methods based on the use of second antibody — antibody against human immunoglobulin — will give high background staining when the target tissue contains human immunoglobulin. This would be manifest when working with B lymphocytes and with tissue bathed in plasma. This technical difficulty can be resolved by direct labeling of the monoclonal antibody.

The different range of antigens recognized by an immune response within the species, in contrast to a response across species (a xenogeneic response), must affect the range of monoclonal antibodies which can be obtained. This difference is important in the context of human monoclonal antibodies for research, diagnostic, or therapeutic purposes. Under normal circumstances, the range of human molecules which can elicit an immune response in man is very limited, whereas many human molecules may be expected to be immunogenic in mice. Autoimmune reactions do occur, and it is possible that in vitro, away from the normal homeostatic mechanisms of the body, autoantibody responses may be induced to a large variety of structures. This is a possibility, as yet not verified experimentally, and the preparation of human monoclonal antibodies may turn out to be possible only against antigens which are polymorphic, or the target of a known autoimmune response.

On the other hand, if we are interested in defining polymorphic antigens, or if we are interested in autoimmunity, immunization within the species offers major advantages. It is difficult to resolve minor structural differences in a xenogeneic immunization, where the recipient is faced with a large number of antigenic differences. In an allogeneic immunization the recipient sees only the polymorphic antigens. Thus, murine monoclonal antibodies have not proved successful in tissue typing (HLA typing); human monoclonal antibodies are more likely to succeed. Where tumor-specific antigen does exist, it is possible that the antigen is a slightly modified normal antigen, which may be recognized within the species, but not in xenogeneic immunization.

## REFERENCES


1. Manser, T., Wysocki, L. J., Gridley, T., Near, R. I., and Gefter, M. L., The molecular evolution of the immune response, *Immunol. Today,* 6, 94, 1985.
2. Goodwin, J. S., OKT3, OKT4, and all that, *J.A.M.A.,* 246, 947, 1981.
3. Zola, H., Speaking personally: monoclonal antibodies as diagnostic reagents, *Pathology,* 17, 53, 1985.
4. Preud'homme, J. L., Les marqueurs lymphocytaires en clinique: interpretation et indications, *Presse Med.,* 14, 669, 1985.
5. Mitchell, M. S. and Oettgen, H. F., *Progress in Cancer Research and Therapy,* Vol. 21, Raven Press, New York, 1982.
6. Boss, B. D., Langman, R., Trowbridge, I., and Dulbecco, R., *Monoclonal Antibodies and Cancer,* Academic Press, Orlando, 1983.
7. Meeker, T. C., Lowder, J., Maloney, D. G., Miller, R. A., Thielemans, K., Warnke, R., and Levy, R., A clinical trial of anti-idiotype therapy for B cell malignancy, *Blood,* 65, 1349, 1985.

8. Meeker, T., Lowder, J., Cleary, M. L., Stewart, S., Warnke, R., Sklar, J., and Levy, R., Emergence of idiotype variants during treatment of B-cell lymphoma with anti-idiotype antibodies, *N. Engl. J. Med.*, p. 312, 1658, 1985.
9. LeBien, T. W., Boue, D. R., Bradley, J. G., and Kersey, J. H., Antibody affinity may influence antigenic modulation of the common acute lymphoblastic leukemia antigen in vitro, *J. Immunol.*, 129, 2287, 1982.
10. Bradwell, A. R., Fairweather, D. S., Dykes, P. W., Keeling, A., Vaughan, A., and Taylor, J., Limiting factors in the localization of tumours with radiolabelled antibodies, *Immunol. Today*, 6, 163, 1985.
11. Kreis, H., Chkoff, H., Vigeral, P., Chatenoud, L., Lacombe, M., Campos, H., Pruna, A., Goldstein, G., Bach, J. F., and Crosnier, J., Prophylactic treatment of allograft recipients with a monoclonal anti-$T3^+$ cell antibody, *Transpl. Proc.*, 17, 1315, 1985.
12. Ortho Multicenter Transplant Study Group, A randomized clinical trial of OKT3 monoclonal antibody for acute rejection of cadaveric renal transplants, *N.Engl. J. Med.*, 313, 337, 1985.
13. Granger, S., Janossy, G., Francis, G., Blacklock, H., Poulter, L. W., and Hoffbrand, A. V., Elimination of T-lymphocytes from human bone marrow with monoclonal T-antibodies and cytolytic complement, *Br. J. Haematol.*, 50, 367, 1982.
14. Jansen, J., Falkenburg, J. H. F., Stepan, D. E., and LeBien, T. W., Removal of neoplastic cells from autologous bone marrow grafts with monoclonal antibodies, *Semin. Hematol.*, 21, 164, 1984.
15. Bast, R. C., De Fabritis, P., Lipton, J., Gelber, R., Maver, C., Nadler, L., Sallan, S., and Ritz, J., Elimination of malignant clonogenic cells from human bone marrow using multiple monoclonal antibodies and complement, *Cancer Res.*, 45, 499, 1985.
16. Treleaven, J. G., Ugelstad, J., Phillips, T., Gibson, F. M., Rembaum, A., Caine, G. D., and Kemshead, J. T., Removal of neuroblastoma cell from bone marrow with monoclonal antibodies conjugated to magnetic microspheres, *Lancet*, i, 70, 1984.

Appendix 1

# SOURCES OF INFORMATION ON EXISTING HYBRIDOMAS

Large numbers of hybridomas have been established in laboratories around the world. Most potential users will prefer to obtain their monoclonal antibodies from another laboratory rather than set up a fusion project. A research team which is already making hybridomas will still find many uses for antibodies made by others. How do you find out what's available? There are two major sources of information, the scientific literature and the commercial suppliers. A list of commercial suppliers of monoclonal antibodies is included in Appendix 2. There are a few additional sources of information which deserve a special mention.

## I. SPECIAL JOURNALS AND HYBRIDOMA LISTINGS

A small number of journals exists largely to publish information about monoclonal antibodies. In addition, public bodies such as the National Institutes of Health publish occasional listings. In view of the large numbers of monoclonal antibodies prepared for diverse uses, these listings are most useful if they focus on a particular area of application. The following is a list of journals and listings which have come to the author's attention:

- *Monoclonal Antibody News:* a forum for publication of the key properties of new monoclonal antibodies in brief form, published by Mary Ann Liebert Inc., 157 East 86th Street, New York
- *Hybridoma:* publishes full-length papers, from the same publisher as *Monoclonal Antibody News*
- *Journal of Immunological Methods;* publishes a special section with summaries of new hybridomas
- *International Cancer Research Data Bank:* publishes occasional compilations of abstracts on specific areas. ICRDB Program Office, National Cancer Institute, Westwood Building, Bethesda, Md., 20205

## II. AMERICAN TYPE CULTURE COLLECTION (ATCC)

This organization maintains a cell bank, and hybridoma lines deposited with the ATCC may be obtained, either frozen or in culture, for the payment of a small fee. This is a very cheap way of obtaining monoclonal antibody. Some of the commercial hybridomas are deposited here, for patent reasons, and are available on condition they are not used for commercial purposes. The catalog is available from American Type Culture Collection, 12301 Parklawn Drive, Rockville, Md., 20852.

Appendix 2

# MANUFACTURERS AND SUPPLIERS OF EQUIPMENT, MATERIALS, AND REAGENTS

This list is not intended as a complete directory of suppliers, but contains the names and addresses of companies supplying specialized materials or apparatus used in the techniques described in this book. The major commercial suppliers of monoclonal antibodies are also listed. For a more complete directory of reagent suppliers, the reader is referred to *Linscott's Directory of Immunological and Biological Reagents,* obtainable from Linscott's, P.O. Box 24, East Grinstead, England. The advertising sections of the major immunological journals, particularly the *Journal of Immunology,* are a good source of information on new products.

The listing is divided into two sections, a directory classified into product categories and an alphabetical listing of companies, with addresses. The latter includes all the companies referred to in the book.

## I. CLASSIFIED DIRECTORY

1. Specialized equipment (general items, such as microscopes, pH meters, centrifuges and balances are not included).
   - Incubators
     - Contherm Scientific
     - Forma
     - National Appliance Co.
   - Laminar Flow and biohazard hoods
     - Gelman
     - Email (Australia)
   - Sterilization equipment
     - Gelman
     - Millipore
     - Sartorius
   - Liquid dispensing equipment
     - Costar
     - Eppendorf
     - Flow Laboratories (Pipetus)
     - Gilson
     - Oxford
     - Gelman
   - Equipment and materials for electrophoresis and chromatography
     - Bio-Rad
     - LKB
     - Pharmacia
     - Behringwerke
     - Shandon Southern
   - ELISA plate Readers
     - BioRad
     - Dynatech
2. Tissue culture plasticware
   - Costar

- Falcon
- Linbro
- Lux
- Nalgene
- Nunc
- Sterilin

3. Media and serum
   - FLOW Laboratories
   - GIBCO
   - OXOID
   - DIFCO
4. General chemicals:
   - BDH
   - Calbiochem-Behring
   - Merck
   - Pierce Chemical Co.
   - Serva
   - SIGMA
   - TAAB
5. Immunological reagents
   - Amersham Corporation
   - Becton Dickinson
   - Bethesda Research Laboratories
   - Bio-Rad
   - Bio-Yeda
   - DAKO
   - DP Diagnostics
   - E-Y Laboratories
   - Janssen Pharmaceutica (colloidal gold reagents)
   - Miles Laboratories
   - SIGMA
   - VECTOR Laboratories
6. Monoclonal antibodies
   - Accurate Chemical
   - AMD (Australian Monoclonal Development
   - Becton Dickinson
   - Boehringer Mannheim
   - BRL
   - Calbiochem-Behring
   - CAPPEL
   - Coulter
   - Dako
   - GIBCO
   - Hybritech
   - Miles Laboratories
   - New England Nuclear (Du Pont)
   - Ortho
   - Sera Lab
   - Serotec

## II. ALPHABETICAL LISTING WITH ADDRESSES

- Accurate Chemical & Scientific Corp., 300 Shames Drive, Westbury, N.Y., 11590, Telex 144617
- Australian Monoclonal Development Pty. Ltd., 65 Dickson Avenue, Artarmon, NSWx 2064, Australia, Telex: AMD AA 73396
- Amersham Corp., White Lion Road, Amersham Buckinghamshire, England, HP7 9LL, Telex 83141 ACTIVAG
- Amicon USA, 182 Conant Street, Danvers, Mass., 01923, Telex 200113
- BDH, Broom Road, Poole BH124NN, England, Telex 41186 or 418123 Tetrag
- Becton Dickinson, 2375 Garcia Avenue, Mountain view, Calif., 94043, Telex 9103382026
- Behringwerke, Postfach 1140, D-3550 Marburg 1
- Bethesda Research Laboratories Inc., P.O. Box 6009, Gaithersburg, Md., 20877, Telex 64210
- Bio-Rad, 2200 Wright Avenue, Richmond, Calif., 94804, Telex 335-358
- Boehringer Mannheim Biochemicals, 7941 Castleway Drive, Indianapolis, Ind., 46250
- BRL-Life Technologies Inc., P.O. Box 6009, Gaithersburg, Md., 20877, Telex 64210
- CSL, 45 Poplar Road, Parkville, Vic. Aust. 3052, Telex AA37050
- Calbiochem-Behring, P.O. Box 12087, San Diego, Calif., 92112, Telex 697934
- Cappel Laboratories Division of Cooper Diagnostics Inc., 1 Technology Court, Malvern, Pa., 19355, Telex 831512
- Contherm Scientific, Box 30-605, Lower Hutt, N.Z., Telex Contherm SeekWN NZ3859
- Corning Medical & Scientific, Corning Glass Works, Medfield, Mass., 02052
- Costar, 205 Broadway, Cambridge, Mass., 02139
- Coulter Immunology, 440 West 20 Street, Hialeah, Fla., 33010
- DP Diagnostics, 16-20 Paget Street, Ridleyton, South Australia, 5008, Telex 658481
- Dakopatts a/s, Produktonsvej 42 P.O. Box 1359 DK 2600 Glostrup, Denmark, Telex 35128
- Damon Corp., 300 Second Avenue, Needham Heights, Mass., 02194
- DIFCO, P.O. Box 1058, Detroit, Mich., 48232, Telex 23-5683
- Eppendorf, P.O. Box 650670 D-2000, Hamburg 65 Feb Rep of Germany, Telex 0217431 5d
- E-Y Laboratories Inc., 127 North Amphlett Blvd., Box 1787, San Mateo, Calif., 94401, Telex 349336
- Falcon: Becton Dickinson Labware, 1950 Williams Drive, Oxnard, Calif., 93030
- Flow Laboratories, 7655 Old Springhouse Road, McLean, Va., 22102
- Forma Scientific, Box 649, Marietta, Ohio, 45750, Telex 24-5394
- Gelman Sciences Inc., 600 South Wagner Road, Ann Arbor, Mich., 48106
- GIBCO-Life Technologies Inc., 421 Merrimack Street, Lawrence, Mass., 01843
- Gilson, 72 rue Gambetta, B.P. 45,95400 Villiers-le-Bel, France, Telex 696682F
- Hybritech Inc., 2945 Science Park Road, La Jolla, Calif., 92037
- Ilford Ltd., Basildon, Essex, U.K.
- Janssen Life Sciences Products, Division of Janssen Pharmaceutica N.V., Turnhoutsenweg 30, B-2340 Beerse, Belgium, Telex 32540
- Kodak: Eastman Kodak Co., 343 State Street, Rochester, N.Y., Telex 97-8481
- Leitz D-6330 Wetzlar, Telex 483849 leiz d

- LKB, Fredsfursstigen 22-24, 5-161 25 Bromma 1, Sweden
- Linbro (Flow Laboratories, Inc.) McLean, Va., 22102
- Lux, see Flow Laboratories
- Merck: E. Merck, AG, Darmstadt, West Germany
- Miles Scientific Div., Miles Labs. Inc., 30W475, North Aurora Road, Naperville, Ill., 60566, Telex 210197
- Millipore Corporation, Ashby Road, Bedford, Mass., 01730, Telex 44-30066
- National Appliance Co, Heinicke Instruments Co., 3000 Taft Street, Hollywood, Fla., 33021, Telex 512610
- Nalgene, Nalge Company, Div. of Sybron Corp., Box 365, Rochester, N.Y., 14602, Telex 97-8242
- New England Nuclear, 549 Albany Street, Boston, Mass., 02118, Telex 940996
- Nunc, P.O. 280, Kamstrup — DK 4000, Roskilde, Denmark, Telex 43115
- Oxford, Lancer, Div. of Sherwood Med., St. Louis, Mo., 63103
- Oxoid Ltd., Basingstoke, Hants, England, RG24 OPW
- Pharmacia Fine Chemicals, Laboratory Separation Division, S-75182 Uppsala, Sweden
- Pierce Chemical Co., P.O. Box 117, Rockford, Ill., 61105, Telex 338534
- Sartorius USA, 026575 Corporate Avenue, Hayward, Calif., 94545
- Sera-Lab Ltd., Crawley Down, Sussex, England RH10 4F, Telex 95317 SERLAB G
- Serotec Ltd., Station Road, Blackthorn, Bicester Oxon OX6 OTP England, Telex 83683
- Shandon Southern Products Ltd., 93-96 Chandwick Road, Astmoor, Industrial Estate, Runcorn, Cheshire, England, WA7 1PK, Telex 627706
- Serva, P.O. Box 105260, D-6900, Heidelberg-1, Telex 461709
- Shandon Southern Products Ltd., 93-96 Chandwick Road, Astmoor
- Industrial Estate, Runcorn, Cheshire, England, WA7 1PK, Telex 627706
- Serva, P.O. Box 105260, D-6900, Heidelberg-1, Telex 461709
- Sigma Chemical Co., P.O. Box 14508, St. Louis, Mo., 63178, Telex 9107610593
- Sterilin, Teddington, Middlesex, England
- TAAB Laboratories Equipment Ltd., 40 Grovelands Road, Reading, Berkes, England
- Union Carbide, Danbury, Conn.
- Vector Laboratories Inc., 1429 Rollins Road, Burlingome, Calif., 94010, Telex 172156
- Wellcome Reagents, 3030 Cornwallis Road, Research Triangle Park, N.C., 27709

Appendix 3

# LABORATORY HAZARDS

All chemicals should be considered toxic. Biological materials carry potential hazards, which may not be known when the materials are first used. Smoking and eating in the laboratory entail the risk of accidental ingestion of toxic materials (in addition to those we are told are already in the food and smoke). Smoking and eating should be banned from the laboratory, as should mouth pipetting. Sensible precautions should be observed with regard to clothing and items, such as books, which pass freely from the laboratory to the office and coffee room. In addition to these general precautions, a number of substances referred to in this book carry special hazards. These have been mentioned in the text and are listed here to serve as a reminder. Note that the failure to list a substance here does not mean it is safe; *all chemicals should be treated as toxic.*

Azide — This forms explosive compounds with metals, especially copper. Check the plumbing and the syringes etc. used in dispensing solutions containing azide.

Picric acid — This is highly explosive when dry. If the rubber bung on the picric acid bottle is caked with dry picric acid, get some advice from a professional chemist before trying to open the bottle.

Dimethylsulfoxide — This solvent has the unusual property of transporting substances dissolved in it through the skin and, thus, requires careful handling.

Acrylamide — This is neurotoxic.

Osmium tetroxide — This is a toxic vapor.

Freund's complete adjuvant — Avoid injecting yourself, as it can cause painful and troublesome local granulomata. The amount injected does not have to be large. If you do inject yourself and suffer no adverse consequences, be especially careful; it is likely that a secondary immunological response will be troublesome even if a first exposure did not lead to a noticeable reaction.

Mice — They bite. More important, the material you wish to inject into the mouse may not be good for you. Apart from Freund's adjuvant, discussed above, biological material may contain viruses etc. The practice of injecting a mouse while someone else holds it is particularly hazardous in this context.

UV light — The UV lamps used to sterilize working areas must be switched off before the work station is used, and a UV-absorbing plastic curtain must be in place at the front of the work station when the lamp is switched on. Unpleasant burns to the eyes and skin are readily produced by these lamps, and the longer-term effects of UV radiation can include carcinogenicity.

Appendix 4

# BUFFERS AND REAGENTS

Most of the buffers used in this book are given in recipe form. However, a few general comments on the making up and use of buffers and reagents may be helpful for some workers whose background training has not included practical biochemistry.

## I. BUFFERS

### A. General Comments

A buffer minimizes pH variations caused by chemical reactions which produce hydrogen or hydroxyl ions. A particular formulation will be effective over a restricted pH range and will not buffer outside this range. When working with living cells or tissue the buffer must also be isotonic; for some biochemical purposes the ionic strength is important. There are particular situations where particular ions have to be avoided, because they interfere with the reaction being carried out. For example, reactions involving the free amino groups of proteins should not be carried out in glycine buffers, since glycine has an amino group. Buffers with N atoms capable of donating electrons interfere in nucleophilic substitutions (Chapter 6, Section V).

### B. Buffer Tables

Tables for the preparation of buffers may be found in:

- Gomori, G., Preparation of buffers for use in enzyme studies, in *Methods in Enzymology,* Vol. 1, Colowick, S. P. and Kaplan, N. O., Eds., Academic Press, New York, 138.
- Dawson, R. M. C., Elliott, D. C., Elliott, W. H., and Jones, K. M., *Data for Biochemical Research,* Oxford University Press, Oxford, England.
- Sober, H. A., Ed., *Handbook of Biochemistry, Selected Data for Molecular Biology,* CRC Press, Boca Raton, Fla., J-227.

### C. Making Up and Keeping

There is no room for flair and panache, or even originality in the preparation of buffers. They must be made up carefully, with a good balance, accurate volumetric apparatus, and a pH meter which is correctly set up. Buffers which have been stored must be checked for microbial growth, and the pH must be adjusted if it has drifted. How this adjustment is made depends on the use of buffer. If pH is important, but ionic strength is not, simply add strong (1 to 10 *N*) HCl or NaOH, dropwise, while stirring and checking pH. If ionic strength is important, the adjustment will have to be made with the appropriate buffer salt at the correct ionic strength (for example, disodium hydrogen phosphate at 0.05 *M* to bring the pH of a 0.05 *M* phosphate buffer up). pH changes slightly with temperature and, also, with dilution.

## II. REAGENTS

"Reagents" is a term covering a wide variety of substances and solutions, and it would not be possible to cover all the things that can go wrong. However, the author has encountered a few misunderstandings and difficulties repeatedly, and these are discussed briefly here.

A. Calculation of Dilutions

If you cannot work out how to make up a 0.1 *M* solution of a substance, given its molecular weight, seek help. However, it is surprising how many competent laboratory workers have trouble calculating further dilutions from a known stock solution. The easy way to do this is using the relation: Volume (V) × concentration (C) is a constant, or, as an equation, $V1 \times C1 = V2 \times C2$.

For example, we need 500 mℓ of 0.05 *M* reagent; the stock solution is 1.5 *M*; how much stock do we dilute? V1 is the unknown; $C1 = 1.5$; $V2 = 500$; $C2 = 0.05$. V1 is calculated to be $500 \times 0.05/1.5$; or 16.67 mℓ.

B. Weighing Out Reagents

If the solid material is stored cold, allow the bottle to reach ambient temperature before opening it, otherwise water will condense from the air onto the chemical, affecting the accuracy and the activity. Use a balance appropriate to the amount to be weighed. Amounts less than 5 mg cannot be weighed accurately without specialized equipment and expertise. If a recipe requires 5 mg in 10 mℓ, it will be a lot quicker to dissolve, say, 5.4 mg in 10.8 mℓ than to fiddle with the amount on the balance until it is precisely 5 mg. However, for the highest precision accurate volumetric glassware must be used, and this restricts flexibility on volumes. The experimenter will usually know whether or not a particular concentration is critical. In this book, if a salt is given without water of crystallization in the recipes, the weight is for anhydrous salts. Hydrates may be used, after making the necessary correction to the weight.

When dissolving up proteins, do not shake the solution vigorously; any froth is a sign of protein denaturation. Some substances can be heated to dissolve them, but biological materials generally cannot.

If the reagent doesn't work as expected, make it up fresh. Check the label on the bottle and the recipe in the paper you are working from. There are many complex reasons for the failure of a reagent, but the reasons are often simple.

# Appendix 5

# UNITS AND PREFIXES

## I. RADIOACTIVITY

The old unit for the quantity of radioactive material was the Curie (Ci). One Curie is the amount of isotope giving $3.7 \times 10^{10}$ disintegrations per second. The new unit is the Becquerel (bq); 1 bq is the amount of material giving one disintegration per second. The Ci is still much loved by recalcitrant old scientists who were not invited to sit on the Nomenclature Committee (including this author). In the quantities used in a biological laboratory, the practical unit is the mCi: 1 mCi = $3.7 \times 10^7$ bq or 37 *M*bq.

## II. PREFIXES AND SYMBOLS FOR MULTIPLES OF UNITS

| Multiple of unit | Prefix | Symbol |
|---|---|---|
| $10^{12}$ | tera | T |
| $10^9$ | giga | G |
| $10^6$ | mega | M |
| $10^3$ | kilo | k |
| $10^2$ | hecto | h |
| 10 | deca | da |
| $10^{-1}$ | deci | d |
| $10^{-2}$ | centi | c |
| $10^{-3}$ | milli | m |
| $10^{-6}$ | micro | $\mu$ (mu) |
| $10^{-9}$ | nano | n |
| $10^{-12}$ | pico | p |
| $10^{-15}$ | femto | f |
| $10^{-18}$ | atto | a |

# INDEX

## A

## D

## G



## H

## I



## K



## L

## M



## N



## O



## P

Q



R



S

## T

U



V



W



X



Y



Z